P. K. Richardson PhD

© P K Richardson 1995
Published by PasTest, Knutsford, Cheshire.

All rights reserved. No part of this publication may be reproduced, stored in a retrieval system, or transmitted, in any form or by any means, electronic, mechanical, photocopying, recording or otherwise without the prior permission of the copyright owner.

First printed 1995

A catalogue record for this book is available from the British Library.

ISBN 0 906896 28 2

Text prepared by Waring Collins Partnership, Ormskirk.

Printed in Great Britain by the Ipswich Book Company, Suffolk

CONTENTS

Introduction v
Preparing for the GMAT vii

Part 1: Mathematics Review 1

1. Arithmetic
    1.1 Basic Terms and Concepts 3
    1.2 Arithmetic Operations 3
    1.3 Basic Properties of Numbers 4
    1.4 Fractions 4
    1.5 Decimals 8
    1.6 Percentages 13
    1.7 Averages 16
    1.8 Powers, Exponents, Roots and Radicals 17

2. Algebra
    2.1 Algebraic Functions 23
    2.2 Problem Solving in Algebra 23
    2.3 Inequalities 25

3. Geometry
    3.1 Geometric Symbols 29
    3.2 Planar Figures 31
    3.3 Perimeters of Planar Figures 36
    3.4 Areas of Planar Figures 37
    3.5 Volumetric (Three Dimensional) Figures 44
    3.6 Co-ordinate Geometry 47

4. Tables and Graphs
    4.1 Tables 51
    4.2 Graphs 52

Part 2: How To Tackle GMAT Maths Questions 57
    Problem Solving (48 questions) 59
    Data Sufficiency (30 questions) 89

Part 3: Four Part Practice Examination 111
    Section 1: Problem Solving 1 (16 questions) 113
    Section 2: Problem Solving 2 (16 questions) 117
    Section 3: Data Sufficiency 1 (20 questions) 121
    Section 4: Data Sufficiency 2 (20 questions) 127
    Answer Key 133
    Self-Scoring Scale 134

Part 4: Analysis of Solutions to Practice Examination 135

INTRODUCTION

The best business schools require applicants to have obtained a high GMAT score. This book has been written for the ambitious GMAT candidate who requires a concise and intensive mathematics revision text.

Many potential MBA candidates do not apply to business schools immediately after graduating from university. In general, business schools prefer candidates who have experience of working in a business environment. If several years have passed since you were called upon to sit an examination, the prospect of having to achieve a high GMAT score, often at short notice, can be daunting. My experience of teaching graduate students in a typical business school, and of directing intensive GMAT revision courses, has convinced me that a book covering the basic maths requirements for GMAT will be invaluable to "rusty" candidates.

Although you may already have a good grasp of the maths principles you need for the GMAT, I would strongly recommend that you quickly refresh your skills by reading this book and tackling some of the practice questions, which are highlighted for convenience. If your maths is rusty, you will need to work through this book systematically and where it becomes apparent that more help is required, this must be sought.

The mathematics questions in the GMAT are not difficult – most are of a comparable standard to Fifth Form maths. Given enough time, most candidates would be able to achieve a high score. However, the greatest problem associated with the GMAT is the time element – your ability to work consistently well under pressure is being tested as well as your mathematical skills. *On average, you have about one minute to answer each of the maths questions in the test.* Since some of the questions are relatively easy, the time saved on these can be spent on the more difficult questions.

The two attributes necessary for you to achieve a high score in the GMAT are technical competence and flexibility. Technical competence in straightforward maths and English is absolutely essential. However, this alone will not guarantee you a high score. The key to turning competence into results is in your ability to be flexible in your approach to each question so that you can find the quickest way to arrive at the correct answer. *The high scores that business schools look for can only be achieved by "smart" candidates, and not by mathematicians or English language specialists.*

This book is split into four parts for ease of use. The first part is a review of the basic mathematics concepts which are important for the GMAT. The points covered have been selected as the most important and relevant to the test. It would be impossible to include every mathematics concept that is taught at school, and most GMAT candidates will not have the time available to work through an exhaustive revision text. By working through this short review systematically, you should quickly become aware of any possible weak areas in your knowledge which require further practice.

Part 2 consists of 48 GMAT standard Problem Solving questions (equivalent to three exam sections) and 30 Data Sufficiency questions, which have been solved in the conventional way and also, where appropriate, by a quicker method (the GMAT method) with a view to illustrating the flexibility of approach that the GMAT requires. The reader is encouraged to find and use the quickest method of solving each question.

Part 3 is a four-part Practice Examination comprising two Problem Solving sections and two Data Sufficiency sections. Part 4 contains the solutions to these questions for self-scoring and analysis.

Over many years I have met hundreds of GMAT candidates and have found that they are generally more concerned about the mathematics questions than about any other aspect of the test. I hope that the advice I have given to the candidates I have met will now be of help to many other MBA candidates.

I am indebted to Jane Bowler and Freydis Campbell, both of PasTest, for their invaluable advice and support during the preparation of this book.

P K Richardson PhD
Research Fellow
Manchester Business School

PREPARING FOR THE GMAT

Taking the GMAT is a stressful and expensive habit! Candidates who approach the test without preparing thoroughly are unlikely to achieve a score that reflects their full potential. Although in theory there is no limit to the number of times you can take the test, remember that each time you ask ETS to send an official report of your score to the business school of your choice, they will send a sheet showing the two most recent scores you have achieved to date. Low scores can seriously affect your chances of acceptance. The following study plan will help you plan your revision time so that you achieve your best possible GMAT score.

Six weeks before the test
- Familiarise yourself with the various types of questions that are used in the test.
- Use the first section of this book to brush up your basic maths. If necessary, do some serious maths revision.
- Work through the second section of this book to learn the smart ways to approach a variety of GMAT maths questions.
- Learn the rules for answering Data Sufficiency questions.
- If you are concerned about your ability to cope with the English questions in the test, use *The Official Guide for GMAT Review* to iron out any problems.

Four weeks before the test
- Sit the Practice Examination in this book under timed conditions. If your score is low, make sure you understand why.
- Sit a complete GMAT Practice Examination using *The Official Guide for GMAT Review* or PasTest's *GMAT Practice Exams* book. You should be aiming for a GMAT score of at least 500.

Two to three weeks before the test
Attend PasTest's one day intensive GMAT preparation course. This includes
- Specific tips and techniques for using your time effectively.
- Strategies on how to approach each type of GMAT question.
- A mock exam which is taken under authentic test conditions.
- Invaluable practice which will give you a competitive edge.
- The chance to meet and talk to other MBA candidates.

Turn to the back of this book for details.

Up to the date of the test
- Attempt two or three more practice examinations using *The Official Guide for GMAT Review*.
- Keep a chart to show how your score has improved.
- **Good Luck!**

ON THE DAY OF THE TEST

Get up in good time and have some breakfast. Give yourself plenty of time to get to the test centre and make sure you know exactly where you are going.

Remember to take
- four sharpened pencils and two erasers
- a silent watch
- your entry form (or completed application form if you are taking the test on stand-by)
- your passport or another document showing your name and photograph
- your credit card or another recognised method of payment if you are taking the test on stand-by
- a drink and snack for the break.

You are not allowed to take
- calculators including watch calculators
- watches with alarms
- paging devices (bleepers)
- rough paper of any kind
- books, dictionaries, pamphlets
- rulers, protractors, compasses, stencils
- slide rules
- highlighter pens.

During the exam
Make sure your desk is well lit and doesn't wobble - after 3 hours these things will start to irritate you.

Mark your answer sheet clearly and rub out any mistakes thoroughly.

Skim through each section before you start so that you don't have any surprises.

Read the questions carefully!

Feel free to make notes and do rough calculations on the *question* book.

Don't spend too much time on any one question. Keep moving.

Don't allow other candidates to worry you during the break. Try not to have post-mortems.

PART I
Mathematics Review

1. ARITHMETIC

1.1 Basic Terms and Concepts

A *whole number* is one which has no fractional part. Numbers such as 1, 5, 100 and 987 are whole numbers.

An *integer* is a whole number. A positive integer is a whole number which is equal to or greater than 1, such as 2, 10 or 57. A negative integer is a whole number which is less than zero such as −1, −10 or −57.

An *odd number* is a number that is not divisible by 2, such as 3, 5, 9 and 65. All odd numbers when divided by 2 leave a remainder of 1.

An *even number* is a whole number that is divisible by 2. Numbers such as 6, 8, 18 and 58 are all divisible by 2 and are therefore even numbers.

A *prime number* is a number that is only divisible by itself or by 1. In other words, a prime number has no factors except 1 and itself. Examples of prime numbers are 5, 7, 11, 23, 29 and 37.

1.2 Arithmetic Operations

Addition − to find the sum of two or more numbers simply add the numbers together. As an example, the sum of 10, 15 and 18 is the same as 10 + 15 + 18 which equals 43.

Subtraction − the difference between two numbers is simply one minus (take away) the other. The difference of 27 and 19 is the same as 27 − 19 which equals 8.

Multiplication − the product of two numbers x and y is $x \times y$ or xy. The product of 8 and 6 is 48.

Division − the quotient of two numbers is the result obtained from dividing one by the other. For example, 48 divided by 8 is written as $48 \div 8$ or $\frac{48}{8}$. The number 48 is referred to as the dividend, 8 as the divisor and the answer 6 as the quotient.

1.3 Basic Properties of Numbers

A *positive number* is one that is greater than zero. It is usually written without a sign. Thus 4 and 57 are positive numbers.

A *negative number* is one that is smaller than zero. Numbers such as −1, −2.5 and −102 are negative numbers.

Operations Involving Signed Numbers

It is important to take note of the sign of a number when carrying out operations such as addition, subtraction, multiplication and division.

Examples:

Subtracting a negative number makes that number positive.
If $x = 4$, $y = −6$ and $z = −3$
then $x − y = 4 − (−6) = 4 + 6 = 10$
$x + y = 4 + (−6) = 4 − 6 = −2$

In operations involving multiplication and division, unlike signs lead to negative products or quotients while like signs yield positive products or quotients.

$x \times y = 4 \times (−6) = −24$
$x \div y = 4 \div (−6) = \frac{4}{−6} = \frac{−2}{3}$
$y \times z = (−6) \times (−3) = 18$
$y \div z = (−6) \div (−3) = 2$

1.4 Fractions

A *fraction* is a part of a unit and has a numerator and a denominator. For the fraction $\frac{3}{4}$, 3 is the numerator and 4 is the denominator. The fraction also denotes a division, in this case 3 is divided by 4.

An *improper fraction* is one in which the numerator is greater than the denominator. As a result, such a fraction can be rewritten as a mixed fraction − one that has a whole number and a fraction.

Examples:

$\frac{8}{5}$ is equivalent to $1\frac{3}{5}$

$\frac{7}{4}$ is equivalent to $1\frac{3}{4}$

Simplifying Fractions

The numerator and denominator of a fraction can be changed without affecting the value of the fraction by multiplying or dividing them by the same number. This is important where it is necessary to simplify a fraction or make the denominator take a particular value.

Examples:

1. $\frac{15}{20}$ is equivalent to $\frac{3}{4}$. This result is obtained by dividing both the numerator and denominator by 5.

2. $\frac{11}{15}$ is the same as $\frac{22}{30}$, obtained by multiplying through by 2.

3. Which of these fractions is the smallest: $\frac{1}{2}$, $\frac{1}{3}$, $\frac{3}{10}$, and $\frac{2}{5}$?

    Solution: Put all the fractions into a form in which they have the same denominator. The lowest common multiple (LCM) or lowest common denominator (LCD) of the denominators is 30. Thus

    $$\frac{1}{2} = \frac{15}{30}, \quad \frac{1}{3} = \frac{10}{30}, \quad \frac{3}{10} = \frac{9}{30}, \text{ and } \frac{2}{5} = \frac{12}{30}$$

    Comparing $\frac{15}{30}$, $\frac{10}{30}$, $\frac{9}{30}$ and $\frac{12}{30}$, it is clear that the smallest is $\frac{9}{30}$ (that is $\frac{3}{10}$).

Adding and Subtracting Fractions

If the denominators of the fractions to be added or subtracted are the same, then the operation boils down to adding or subtracting the numerators.

Examples:

$$\frac{5}{6} + \frac{2}{6} \text{ equals } \frac{7}{6}$$

$$\frac{6}{8} + \frac{3}{8} \text{ equals } \frac{9}{8}$$

$$\frac{7}{8} - \frac{3}{8} \text{ equals } \frac{4}{8} \text{ or } \frac{1}{2}$$

$$\frac{7}{9} - \frac{2}{9} - \frac{3}{9} \text{ equals } \frac{2}{9}$$

Where the denominators are different, it is necessary to convert the fractions to the same denominator (the LCM or LCD). As mentioned above, the LCD is the lowest number that can be divided evenly by all the given denominators.

$$\text{The LCD of } \frac{1}{3} \text{ and } \frac{1}{2} \text{ is } 6$$

$$\text{The LCD of } \frac{1}{15} \text{ and } \frac{3}{4} \text{ is } 60$$

Examples:

1. Add $\frac{2}{7}$ and $\frac{3}{5}$

    Solution: The LCD of 7 and 5 is 35 thus

    $$\frac{2}{7} + \frac{3}{5} \text{ is equal to } \frac{10}{35} + \frac{21}{35}$$

    $$\text{or } \frac{10 + 21}{35} \text{ or } \frac{31}{35}$$

2. Subtract $\frac{2}{5}$ from $\frac{7}{8}$

Solution: The LCD of 8 and 5 is 40 thus

$$\frac{7}{8} - \frac{2}{5} \text{ equals } \frac{35-16}{40}$$

or $\frac{19}{40}$

The whole idea underlying these operations is to convert the fractions so that they have the same denominator and then add, or subtract, the numerators as the case may be.

Multiplying and Dividing Fractions

To multiply or divide fractions, they need not have the same denominator. However all mixed fractions must first be converted to improper fractions.

Examples:

$$\frac{3}{8} \times \frac{4}{5} \text{ equals } \frac{3 \times 4}{8 \times 5} = \frac{12}{40} \text{ or } \frac{3}{10}$$

$$1\frac{3}{4} \times \frac{3}{8} \text{ equals } \frac{7}{4} \times \frac{3}{8} = \frac{21}{32}$$

With fractions (and percentages) the term 'of' means multiplication. Thus, $\frac{1}{4}$ of 40 oranges is equal to $\frac{1}{4} \times 40 = 10$ oranges.

To divide fractions, it is necessary to convert all mixed fractions to improper fractions, invert the second fraction and then multiply them together.

Examples:

$$\frac{3}{8} \div \frac{1}{4} \text{ equals } \frac{3}{8} \times \frac{4}{1} = \frac{12}{8} = \frac{3}{2}$$

$$\frac{1}{4} \div \frac{2}{3} \text{ equals } \frac{1}{4} \times \frac{3}{2} = \frac{3}{8}$$

$$4\frac{1}{2} \div 1\frac{2}{3} \text{ equals } \frac{9}{2} \times \frac{3}{5} = \frac{27}{10}$$

Practice Problems Involving Fractions

1. Change $\frac{19}{6}$ to a mixed number.
2. Reduce $\frac{36}{108}$ to the lowest possible terms.
3. Add $1\frac{1}{2}$, $2\frac{1}{3}$ and $\frac{5}{6}$.
4. Subtract $2\frac{1}{4}$ from 6.
5. Which of the following is the largest: $\frac{5}{6}$, $\frac{2}{3}$, $\frac{3}{4}$ and $\frac{7}{12}$?
6. The cost of a man's car is $\frac{3}{7}$ of his annual income. If his income is $14,700 what is the cost of his car?

Answers: (1) $3\frac{1}{6}$ (2) $\frac{1}{3}$ (3) $\frac{14}{3}$ (4) $3\frac{3}{4}$ (5) $\frac{5}{6}$ (6) $6300

1.5 Decimals

A *decimal* (a number with a decimal point) is actually a fraction, the denominator of which is understood to be 10 or some power of 10. The number of digits (referred to as places) after the decimal point determines the power of the denominator.

Fractions can be converted into decimals and vice versa. A fraction can be converted into decimals by dividing the numerator by the denominator.

Examples:

0.7 is equivalent to $\frac{7}{10}$

0.67 is equivalent to $\frac{67}{100}$

0.625 is equivalent to $\frac{625}{1000}$

$\frac{3}{4}$ is 3 divided by 4 which equals 0.75

$\frac{3}{8}$ is equal to 0.375

$\frac{3}{25}$ is equal to 0.12

For the GMAT candidates are expected to know the following fractions, their decimals and percentages:

A half = $\frac{1}{2}$ = 0.5 = 50 per cent

A quarter = $\frac{1}{4}$ = 0.25 = 25 per cent

An eighth = $\frac{1}{8}$ = 0.125 = 12.5 per cent

Operations With Decimals

Addition and subtraction of decimals are done exactly as with whole numbers provided the decimal points are kept in a vertical line, one under the other.

Examples:

1. Add 7.43 and 2.481

   Solution: 7.43
              + 2.481
              ───────
                9.911

2. Subtract 3.57 from 7.43

   Solution: 7.43
              − 3.57
              ──────
                3.86

Decimals are multiplied in the same way as whole numbers. The number of decimal places in the product is equal to the sum of the number of decimal places in the numbers that are being multiplied together.

Examples:

1. Multiply 2.5 by 0.5

   Solution:
       2.5
   x 0.5

   1.25

2. Multiply 2.52 by 3.61

   Solution: 2.52 (2 decimal places)
              x 3.61 (2 decimal places)

                252
             15120
             75600

          9.0972 (4 decimal places)

Four slightly different techniques are involved in the division of decimals.

(a) Where the dividend only is a decimal, the division is the same as that of whole numbers.

Example:

        Divide 9.18 by 3

$$\begin{array}{r} 3.06 \\ 3\overline{)9.18} \\ \underline{9} \\ 01 \\ \underline{0} \\ 18 \\ \underline{18} \end{array}$$

        Solution: 3.06

(b) Where the divisor only is a decimal, the decimal point of the divisor is omitted and as many zeros as there are places in the divisor are placed to the right of the dividend.

Example:

        Divide 1050 by 3.5

        Solution: $\frac{1050}{3.5}$ equals $\frac{10500}{35}$ or 300

(c) Where both dividend and divisor are decimals, the decimal point in the divisor is omitted and the decimal point in the dividend is moved to the right by the same number of places as there are in the divisor. If there are not enough places in the dividend, zeros must be added to make up the difference.

Example:

Divide 6.42 by 0.321

Solution: $\frac{6.42}{0.321}$ equals $\frac{6420}{321}$ or 20.

(d) Where neither the divisor nor the dividend is a decimal, division will lead to a decimal if the divisor is larger than the dividend.

Examples:

1. Divide 3 by 5

$$\frac{3}{5} = \frac{6}{10} = 0.6.$$

2. Divide 12 by 16

$$\frac{12}{16} = \frac{3}{4} = 0.75.$$

Practice Problems Involving Decimals

1. How much is $5.00 + $4.34 + $22.03 + $3.16?

2. John bought 5 packets of crisps for 15 cents each, 2 bars of chocolate for 35 cents each and 2 drinks for 20 cents each. How much change did he get from $2?

3. Evaluate 5.049 − 2.567.

4. Evaluate 27.75 ÷ 0.25.

5. Multiply 6.625 by 0.3.

Answers: (1) 34.53 (2) 15 cents (3) 2.482 (4) 111.0 (5) 1.9875

1.6 Percentages

The symbol % (per cent) stands for part of a hundred. Thus, 30 per cent means 30 out of 100 or $^{30}/_{100}$ or 0.3. Thus fractions, decimals and percentages are interchangeable.

Examples:

1. Write 25% as a fraction and as a decimal.

    Solution: $25\% = \frac{25}{100} = \frac{1}{4}$ or 0.25

2. Find $12\frac{1}{2}\%$ of 72.

    Solution: $12\frac{1}{2}\%$ of $72 = \frac{12\frac{1}{2}}{100} \times 72 = \frac{1}{8} \times 72 = 9$

3. If there are 18 women in a class of 63, what is the percentage of men in the class?

    Solution: Number of men in class = 63 − 18 = 45

    Percentage of men = $\frac{45}{63} \times 100 = 71\frac{3}{7}\%$

In per cent problems, the whole is 100%. Thus per cent changes, per cent increases or per cent decreases are special problems which require the use of the right numbers to achieve the right solutions. The formula that applies is

$$\text{Per cent change} = \frac{(\text{new amount}) - (\text{original amount})}{(\text{original amount})} \times 100$$

Where the new amount is greater than the original, the numerator of the formula will be positive and the result will be a per cent increase. Conversely, where the new amount is less than the original the numerator will be negative and the result will be a per cent decrease.

If a problem involves an increase in a quantity by x per cent (10% say) then the new amount is $(100+x)$ per cent (110%) of the original.

Example:

If the price of a car rises by 10% in a year, what is the new price if the original price was $20,000?

Solution: New price $= (100 + 10)\%$ of $20,000

$$= \frac{110}{100} \times \$20,000$$

$$= \$22,000$$

Similarly if a problem involves a percentage decrease of x per cent (10% say) then the new quantity is $(100 - x)$ per cent (90%) of the original.

Example:

The tax on an item in 1980 was $20,000. If the tax was reduced by 5 per cent in 1981, how much tax had to be paid in 1981?

Solution: New tax $= \frac{95}{100} \times \$20,000$

$$= \$19000$$

Practice Problems Involving Percentages

1. What is $33\frac{1}{3}\%$ of $\frac{1}{3}$ of 27?

2. A mixture of water and sand weighs 120 lbs. What percentage of the weight of the mixture is water if the water weighs 30 lbs?

3. The price of petrol in 1990 was $24 per barrel as against $30 in 1980. What was the percentage decrease in price in the 1980-90 period?

4. A money box contains 400 coins. Of these 20 per cent are dimes, 30 per cent are nickels and the rest are quarters. What amount of money is in the box?

(1 dime = 10 cents), (1 nickel = 5 cents), (1 quarter = 25 cents)

Answers:
1. $\dfrac{33\frac{1}{3}}{100} \times \dfrac{1}{3} \times 27$

   $= \dfrac{1}{3} \times \dfrac{1}{3} \times 27$

   $= 3$

2. Total weight of mixture = 120 lbs
   Weight of water = 30 lbs

   Percentage of water = $\dfrac{30}{120} \times 100$

   = 25 per cent

3. Final price = $24
   Original price = $30

   Percentage change = $\dfrac{24-30}{30} = \dfrac{-6}{30} = \dfrac{-1}{5} = -20\%$

4. Amount of dimes   = 0.1 × 80 = 8
   Amount of nickels   = 0.05 × 120 = 6
   Amount of quarters = 0.25 × 200 = 50
   Amount of money   = 8 + 6 + 50 = $64

1.7 Averages

In the GMAT averages refer to arithmetic means, defined as the ratio of the sum of n numbers to n.

Example:

The sum of 4 numbers is 6+9+10+15 which equals 40.
Average = $\frac{40}{4}$ = 10

Practice Problems Involving Averages

1. What is the average mark of a student who scored 75 in English, 65 in Maths, 70 in French and 30 in Art?

2. John drove at 50 mph from Richmond to Manchester and then at 60 mph from Manchester to Boston. If the distance between Richmond and Manchester is 200 miles and that from Manchester to Boston is 120 miles, what was his average speed for the whole journey?

Answers:

1. Average mark = $\frac{75+65+70+30}{4}$

   = 60.

2. Time taken from Richmond to Manchester = $\frac{200}{50}$ = 4 hours

   Time taken from Manchester to Boston = $\frac{120}{60}$ = 2 hours

   Total time taken = 4 + 2 = 6 hours
   Total distance covered = 200 + 120 = 320

   Average speed = $\frac{\text{total distance}}{\text{total time}}$

   = $\frac{320}{6}$ = 53.3 mph

1.8 Powers, Exponents, Roots and Radicals

Powers and Exponents

When two or more numbers are multiplied together, the result is called the *product* and the numbers are called the *factors* of that particular product. For example, for 6 × 8 = 48, 48 is the product and 6 and 8 are the factors of 48.

Where the factors of a number are the same, an *exponent* (also called a *power*) may be used to indicate the number of times the factor (the base) appears.

Examples:

$$6 \times 2 \times 6 = 6^2 \times 2 = 72 \text{ (I)}$$
$$5 \times 5 \times 5 = 5^3 = 125 \text{ (II)}$$

In (I) the exponent of 6 is 2.
In (II) the exponent of 5 is 3.

If the exponent is 2, we say the base is *squared*. If the exponent is 3, we say the base is *cubed*. A *perfect square* is a number that results when a whole number is squared, such as 4 (2^2), 9(3^2) and 25(5^2).

Square Roots and Cube Roots

If a number has two equal factors, each factor is called a *square root* of the number.
For example 25 = 5 × 5, that is, 5 is the square root of 25.
This is written as $\sqrt{25} = 5$.

Since 64 = 4 × 4 × 4, then 4 is the *cube root* of 64. This is written as $\sqrt[3]{64} = 4$.

Square and cube roots are not always whole numbers.
For example $\sqrt{3} = 1.732$

In the GMAT it is uncommon for candidates to be asked to compute the square (or cube) root of numbers that are not perfect squares (or cubes). The two most common exceptions are $\sqrt{2}$ and $\sqrt{3}$ which equal 1.414 and 1.732 respectively.

Quite frequently in the GMAT candidates have to determine the square root of non-perfect square numbers in terms of $\sqrt{2}$ and $\sqrt{3}$. These simplifications are easily done by factorising out any perfect square factors which are subsequently evaluated.

Examples:

1. Simplify $\sqrt{50}$

    Solution: We need to recognise that $\sqrt{50}$ is the same as $\sqrt{25 \times 2}$

    Thus $\sqrt{50} = \sqrt{25 \times 2}$

    $= \sqrt{25} \times \sqrt{2}$

    $= 5\sqrt{2}$

2. In a right-angled isosceles triangle, AB and BC (the two sides including the right angle) are 8 inches long. What is the length of the hypotenuse?

    A: 8   B: $6\sqrt{2}$ C: $8\sqrt{2}$ D: $8\sqrt{3}$ E: $6\sqrt{3}$

    Solution:

    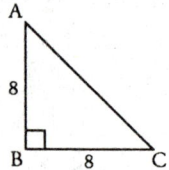

    $AC^2 = 8^2 + 8^2$ (Pythagoras Theorem)
    $AC = \sqrt{64 + 64}$
    $= \sqrt{64 \times 2}$
    $= \sqrt{64} \times \sqrt{2}$
    $= 8\sqrt{2}$

    The correct answer is (C).

Radicals

Radicals are numbers like $\sqrt{2}, \sqrt{3}$ and $\sqrt{5}$. The conditions under which the operations of addition, subtraction, multiplication and division are performed on or with radicals are the same as those for letters in an algebraic expression. The radical acts as a unit.

Addition and subtraction of radicals is possible if the radicals are the same, in which case we only add or subtract their coefficients.

Examples:

Simplify the following

1. $\sqrt{3} + \sqrt{2}$

    Solution: This cannot be simplified further.

2. $3\sqrt{2} + 4\sqrt{2}$

    Solution: $3\sqrt{2} + 4\sqrt{2} = \sqrt{2}(3+4)$
    $= 7\sqrt{2}$

3. $\sqrt{12} + 3\sqrt{3}$

    Solution: $\sqrt{12} + 3\sqrt{3} = \sqrt{4 \times 3} + 3\sqrt{3}$
    $= 2\sqrt{3} + 3\sqrt{3}$
    $= \sqrt{3}(2+3)$
    $= 5\sqrt{3}$

4. $\sqrt{125} - \sqrt{20}$

Solution: $\sqrt{125} - \sqrt{20} = \sqrt{25 \times 5} - \sqrt{5 \times 4}$
$= \sqrt{25} \times \sqrt{5} - \sqrt{4} \times \sqrt{5}$
$= 5\sqrt{5} - 2\sqrt{5}$
$= \sqrt{5}\,(5-2)$
$= 3\sqrt{5}$

Note that in simplifying radicals which contain more than one term under the radical sign, we must combine the terms first before taking the square root.

Examples:

Simplify $\sqrt{7+9} = \sqrt{16} = 4$ (and not $\sqrt{7} + \sqrt{9}$)

Simplify $\sqrt{\frac{3x^2}{16} + \frac{x^2}{16}}$

$= \sqrt{\frac{3x^2 + x^2}{16}} = \sqrt{\frac{4x^2}{16}} = \frac{2x}{4} = \frac{x}{2}$

Multiplication and division of radicals is much simpler and follows the rules of normal algebra.

Examples:

1. $\sqrt{2} \times \sqrt{3} = \sqrt{2 \times 3} = \sqrt{6}$
2. $\sqrt{3} \times \sqrt{3} \times \sqrt{3} = (\sqrt{3})^3 = 3\sqrt{3}$
3. $\sqrt{3} \div \sqrt{2} = \sqrt{3}/\sqrt{2} = \sqrt{\frac{3}{2}}$
4. $\sqrt{7+5} \div \sqrt{1+2} = (\sqrt{12}) \div \sqrt{3}$
$= \sqrt{4 \times 3} \div \sqrt{3}$
$= 2\sqrt{3} \div \sqrt{3}$
$= \frac{2\sqrt{3}}{\sqrt{3}}$
$= 2$

Practice Problems Involving Radicals

Simplify

(1) $\sqrt{18}$ (2) $\sqrt{125}$ (3) $\sqrt{42-6}$ (4) $\sqrt{75} - \sqrt{12}$

Answers: (1) $3\sqrt{2}$ (2) $5\sqrt{5}$ (3) 6 (4) $3\sqrt{3}$

Powers and Exponents - Basic Rules

The following basic rules apply when dealing with exponents and are useful especially when more complex items are included.

Rule 1: $x^a \times x^b = x^{a+b}$

Example: $3^3 \times 3^2 = \underline{3 \times 3 \times 3} \times \underline{3 \times 3} = 3^5$

Rule 2: $(x^a)^b = x^{a \times b} = x^{ab}$

Example: $(3^4)^2 = (3 \times 3 \times 3 \times 3)^2 = 3^{4 \times 2} = 3^8$

Rule 3: $x^a \div x^b = x^{a-b}$

Example: $3^4 \div 3^2 = 3 \times 3 \times 3 \times 3 / 3 \times 3 = 3 \times 3 = 3^2$
or 3^{4-2} or 3^2

Rule 4: $(xy)^a = x^a y^a$

Rule 5: $\left(\dfrac{x}{y}\right)^a = \dfrac{x^a}{y^a}$

Convention 6: $\dfrac{1}{x} = x^{-1}$

$\dfrac{1}{x^a} = x^{-a}$

Practice Problems Involving Powers and Exponents

Evaluate the following:

1. $2^3 \times 4^2$

2. $\dfrac{2.5 \times 10^4}{1.25 \times 10^2}$

3. $\dfrac{6^4}{3^3}$

4. $x^2 y^3 \div x^3 y$

5. $3^{-3} \times 9$

6. $4^3 \div 2^3$

7. If $\dfrac{3.64 \times 10^n}{1.82} = 200$, what is n?

Answers: (1) 128   (2) 200   (3) 48   (4) $\dfrac{y^2}{x}$

                (5) $\dfrac{1}{3}$ (6) 8   (7) 2

2. ALGEBRA

The word algebra is derived from an Arabic term which means a reunion of broken parts. Algebra is a mathematical system that uses variables to generalise mathematical operations.

In algebraic problems we are required to determine the value of an unknown quantity usually represented by a letter such as x, y or z. Algebraic problems may involve simplifications of fractions, derivation of equations involving one or more unknowns and the solution of equations to determine the value of the unknown quantity or quantities.

2.1 Algebraic Functions

Simplifications of algebraic functions follow the basic rules of arithmetic.

Examples:

1. $\dfrac{2x}{y} + \dfrac{1}{y} = \dfrac{2x + 1}{y}$

2. $\dfrac{2a}{b} + \dfrac{3b}{a} = \dfrac{2a^2 + 3b^2}{ba}$

3. $\dfrac{3x^2}{x} = 3x$

4. $\dfrac{x}{6} + \dfrac{2x}{5} = \dfrac{5x + 12x}{30} = \dfrac{17x}{30}$

2.2 Problem Solving in Algebra

The key to solving algebraic problems is to understand clearly what is asked and to derive the correct algebraic expression or equation or inequality. The correct solution of the expression follows the basic rules of arithmetic.

In solving equations, it is important to note that the validity of the equation is not jeopardised as long as the same mathematical operation is performed on both sides of the equation.
For example the equation $2x + 6 = 10$ remains the same if the number 10 is added to (or subtracted from) both sides of the equation.
Thus $2x + 6 = 10$ is the same as $(2x + 6) + 10 = (10) + 10$.

Examples of Algebraic Problems

1. What is the largest of three consecutive positive integers if they sum to 54?

    Solution: Let the largest integer be x then the other two are $(x-1)$ and $(x-2)$

    Thus $(x-2) + (x-1) + x = 54$
    $3x - 3 = 54$
    $3x = 57$
    and $x = 19$

    The integers are 17, 18 and 19.

2. A sociology class has 40 men and women. If there are 8 more men than women, how many women are in the class?

    Solution: Let W be number of women and M be number of men in the class.

    Then $W + M = 40$ (1)
    Also $W + 8 = M$ (2)

    Solving these simultaneously by substituting (2) into (1), we have

    $W + (W+8) = 40$
    or $2W = 32$
    giving $W = 16$

3. A shop sells two types of ice cream, vanilla and strawberry. The price of vanilla ice cream is $4 and that of strawberry $3. On a particular day the shop made total sales of $102. If 32 ice creams were sold that day, how many were vanilla?

Solution: Let the number of vanilla ice creams sold be x. Then the number of strawberry ice creams sold is $(32-x)$.
Thus sales $102 = 3(32 - x) + 4x$.
$x = 6$

2.3 Inequalities

In manipulating inequalities, you can add to or subtract from both sides of the inequality without affecting the validity of the inequality.

Example:

$a > b$
$a + (3) > b + (3)$, adding 3 to both sides
$a - 5 > b - 5$, subtracting 5 from both sides

You can also multiply or divide both sides of the inequality by any **positive** number without affecting the validity of the inequality.

Example:

$a > b$
$a \times 3 > b \times 3$ (multiplying by 3)
or $\dfrac{a}{6} > \dfrac{b}{6}$ (dividing through by 6)

But if you multiply or divide the inequality by a **negative** number, the direction of the inequality is reversed.

Example:

$a > b$
$a \times (-3) < b \times (-3)$, multiplying by (-3)
or $\dfrac{a}{(-5)} < \dfrac{b}{(-5)}$ (dividing through by -5)

Solving Equations and Inequalities

In solving equations (or inequalities) the first step is to determine what quantity or letter has to be isolated. The next steps involve getting this to one side of the equation with all other terms on the other side. This is referred to as making that particular quantity or letter the subject of the relationship.

Examples:

1. Solve $5x + 3 = 18$

    Solution: $5x + 3 = 18$
    $5x = 15$ (subtract 3 from both sides)
    $x = 3$ (divide both sides by 5)

2. If $ax + 6 = 3x + 6 + K$, what is x?

    Solution: $ax - 3x + 6 = 6 + K$ (subtract $3x$ from both sides)
    $ax - 3x = K$ (subtract 6 from both sides)
    $x(a-3) = K$ (factorise left hand side)
    $x = \dfrac{K}{(a-3)}$ (dividing through by $(a-3)$)

Further Examples of Algebraic Problems

1. Three consecutive even integers have a sum of 36. What is the first number?

    Solution: Let the numbers be x, $x + 2$ and $x + 4$
    Then $x + (x+2) + x + 4 = 36$
    $3x + 6 = 36$
    $3x = 30$
    $x = 10$

2. What principal invested at $12\frac{1}{2}$ per cent simple interest for 6 years yields $250?

Solution: Let the principal be P.
Then interest = P × $\frac{12.5}{100}$ × 6 = 250

∴ P = $\frac{250}{0.125 \times 6}$ = $333

3. A man travels from A to C via B. If he travels the distance AB at 60 mph taking 3 hours and BC at 50 mph taking 2 hours, what is the average speed for the whole journey?

Solution: Let the average speed be *x* mph
Then $x = \frac{\text{total distance}}{\text{total time}}$
Total time = 3 + 2 = 5 hours
Total distance = AB + BC
= 60 × 3 + 50 × 2 = 280

∴ $x = \frac{280}{5}$ = 56 mph

4. A GMAT class has 30 students. If there are 14 more men than women, how many men are in the class?

Solution: This is a problem that involves the solution of simultaneous equations.

If we let M equal number of men and W number of women, then
M + W = 30 (1)
Also M = 14 + W (2)

From (1) W = 30 − M and substituting this into (2) we obtain
M = 14 + (30−M)
giving M = 22

27

Practice Problems Involving Algebra

1. John can mow a lawn in 45 minutes while Ben takes twice as much time to do the same job. If they work together, how long will it take to mow the lawn?

2. A man's age is 4 times that of his son. In 20 years time the man's age will be double that of his son. How old is the man now?

3. The formula for determining the velocity of a body dropped from a height h feet is given by $V = \sqrt{2gh}$. Make h the subject of the formula.

Answers: (1) 30 minutes   (2) 40   (3) $h = \dfrac{v^2}{2g}$

3. GEOMETRY

3.1 Geometric Symbols

The most common geometric symbols used in GMAT problems are for lines, circles and angles.

Lines may be represented as $\overline{AB}$ or A•———•B (line AB).

Perpendicular lines (two lines at right angles to each other) may be represented by ⊥.

Parallel lines are represented by ||.

Circles are represented by ⊙.

Angles are commonly represented in three ways:

a) by a small letter written within the angle such as ⦣x⟩
b) by a capital letter at its vertex such as ∠B
c) by three capital letters, the middle letter indicating the vertex such as ∠ABC.

An angle is a figure formed by two lines meeting at a point. Angles may be
acute (less than 90°)
right angle (equal to 90°)
obtuse (greater than 90° but less than 180°) or
straight (equal to 180°).

These may be shown as follows:

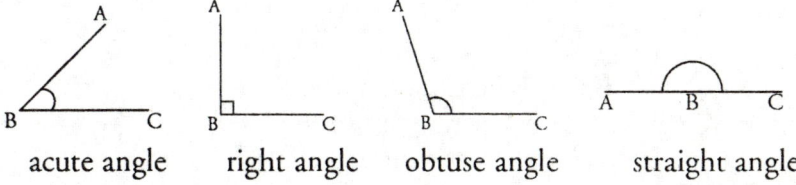

acute angle   right angle   obtuse angle   straight angle

In the above angles, B is called the *vertex* of the angle and the sides AB and BC are the *sides* of the angle.

Two angles are *complementary* if they sum to 90°.
Two angles are *supplementary* if they sum to 180°.

Example:

What is the complement of 60°?

Solution: Let the complement of 60° be *x*
Then *x* + 60 = 90°
Thus *x* = 30°

Two lines intersect if they cross each other. If they cross each other at 90° then they are said to be perpendicular to each other.

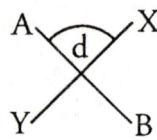 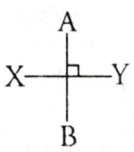

AB intersects XY at d°   AB is perpendicular to XY

Two lines which can never intersect each other are said to be parallel.

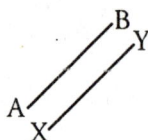

AB is parallel to XY, written as AB || XY

Angles formed by a line that crosses parallel lines

If AB and XY are parallel to each other and CD is a straight line crossing them as in the diagram, then

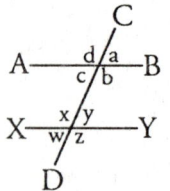

d = b (vertically opposite angles)
c = w (corresponding angles)
d = x (corresponding angles)
c = y (alternate angles)
x = b (alternate angles)

3.2 Planar Figures

Triangles

A triangle is a closed three-sided figure. The following figures are all triangles

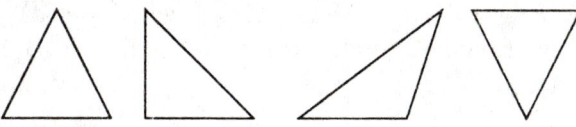

The sum of the three angles in a triangle is 180°.

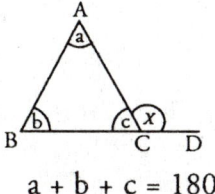

$a + b + c = 180$

In the figure above, $x = a + b$, expressed in a theorem that the exterior angle of a triangle (x) equals the sum of the two interior opposite angles, $(a + b)$.

Isosceles Triangle

A triangle which has two sides equal is referred to as an isosceles triangle. In such a triangle, the angles opposite the equal sides are also equal.

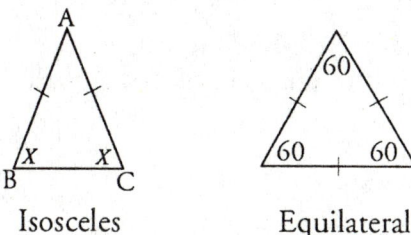

Isosceles   Equilateral

Equilateral Triangle

A triangle with all three sides equal is called an equilateral triangle. All its three angles are also equal, being 60° each.

Right (Angled) Triangle

A triangle which has a right angle (90°) is called a right (angled) triangle. Since the remaining two angles sum to 90°, they are complementary.

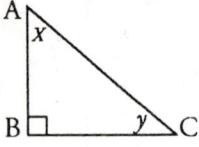

$$x + y = 90°$$

In a right triangle, the side opposite the right angle is called the hypotenuse. The relationship between the sides of a right triangle is expressed by the Theorem of Pythagoras which states that:

The square of the hypotenuse is equal to the sum of the squares of the other two sides.

Using the triangle ABC above, the Theorem states that
$AC^2 = AB^2 + BC^2$

Example:

If the hypotenuse of a right triangle is 5 m and one side is 4 m, what is the length of the third side?

Solution: Let the length of the third side be x, then from the Theorem of Pythagoras,

$$5^2 = 4^2 + x^2$$
$$x^2 = 5^2 - 4^2 = 9$$
$$x = 3 \text{ m}$$

Two Special Right Triangles

The angles, sides and relationships between the sides of these two special right triangles are worth noting, since familiarity with them will save time on the GMAT.

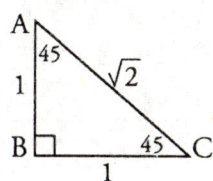

The Isosceles Right Triangle   The 30-60° Right Triangle

Examples:

1. The two equal sides of an isosceles right triangle measure 10 in. What is the length of the hypotenuse?

    Solution: Notice that the ratio of the side to the hypotenuse is $1:\sqrt{2}$.

    Therefore if the side is 10 in, then the hypotenuse must be $10\sqrt{2}$.

2. The hypotenuse of a 30-60° right triangle is 40 in. What are the lengths of the other two sides?

    Solution: The ratio of the hypotenuse to the side opposite the 30° angle and to the side opposite the 60° angle is $2:1:\sqrt{3}$.

    Therefore if the hypotenuse is 40 in, then the sides are respectively 20 and $20\sqrt{3}$ in.

Quadrilaterals

A quadrilateral is a closed four-sided figure. The following figures are all quadrilaterals.

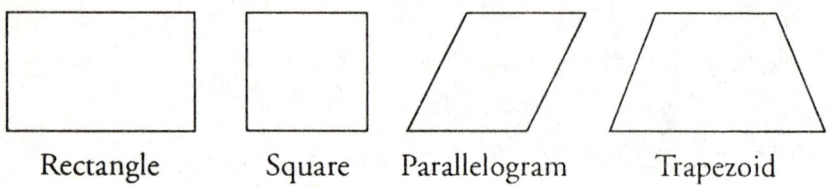

Rectangle   Square   Parallelogram   Trapezoid

Theorem: The sum of the four angles in a quadrilateral is 360°.

A *rectangle* is a quadrilateral with all four angles equal to 90°. The long sides are called the lengths and the short sides are called widths or breadths.

A *square* is a rectangle with all four sides equal.

A *parallelogram* is a quadrilateral in which both pairs of opposite sides are parallel.

A *trapezoid* is a quadrilateral with two sides parallel.

A *rhombus* is a parallelogram with all four sides equal.

Circles

A circle is the locus of all points equidistant from a point called the centre.

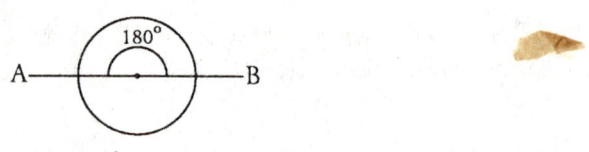

A circle contains 360°

In the figure above, AB divides the circle into two semi circles.

Referring to the figure below,

The *circumference* of a circle is the curved line bounding the circle.

An *arc* of a circle is any part of the circumference.

A *chord* is a line segment connecting any two points on the circle.

A *radius* of a circle is a line segment joining the centre of the circle to any point on the circumference.

A *diameter* is a chord that passes through the centre of the circle.

A *tangent* to a circle is a line that touches the circle at only one point.

A *secant* is a chord that extends in either one or both directions.

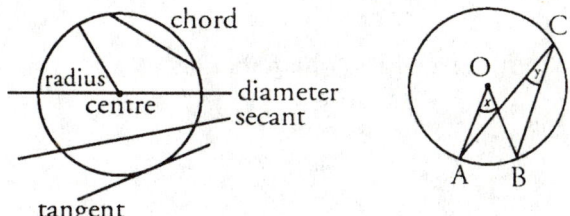

A *central angle* such as x in the diagram above is an angle subtended at the centre of the circle by an arc or chord.

An *inscribed angle* such as y is the angle subtended on any part of the circumference by a chord or arc of a circle.

3.3 Perimeters of Planar Figures

The *perimeter of a planar figure* is the distance around that figure.

Example:

Find the perimeter of the figure below.

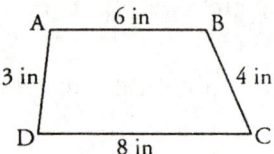

Solution: The perimeter of the figure ABCD is

3 + 6 + 4 + 8 = 21 in.

The *perimeter of a square* is four times the length of one side.

Example:

Find the perimeter of the square below.

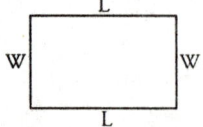

Perimeter = 4 X 4 = 16 in.

The *perimeter of a rectangle* is equal to twice the sum of the length and width.

In the figure below, the perimeter equals 2 (L+W).

The *perimeter of a circle* (referred to as the *circumference*) is equal to the product of the diameter and π, (πd) or $2\pi r$, where d is the diameter, r is the radius and π is $\frac{22}{7}$ (or 3.14). In the GMAT answers are usually required in terms of π.

Example:

What is the circumference of a circle with a radius of 4 cm?

Solution: Circumference = $2\pi r$
= 2 X π X 4
= 8 π cm.

3.4 Areas of Planar Figures

The *area of a planar figure* is the total space within the figure. It is expressed in square denominations such as 'square inches', 'square feet' and 'square metres'. It is absolutely essential to express all dimensions in the same denomination (units) before computing the area.

Area of Rectangles and Squares

The area of a rectangle is the product of the length and the width.
A = L X W.

Example:

Find the area of the rectangle below.

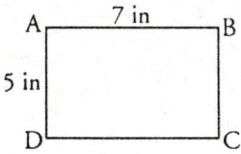

Solution: Area of ABCD = 5 X 7 = 35 sq in.

Since a square is a rectangle with equal sides, the area of a square is the square of one side.
Area of square = L^2

Example:

What is the area of a square with sides 10 cm?

Solution: Area = 10 X 10 = 100 cm²

Area of a Parallelogram

The area of a parallelogram is given by the product of its base and height.
Referring to the diagram below, area = b X h.

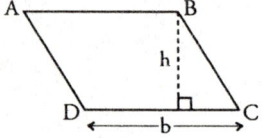

Example:

In the previous diagram, what is the area of parallelogram ABCD if b is 10 cm and h is 7 cm?

Solution: Area = 7 X 10 = 70 sq cm.

Area of a Trapezoid

The area of a trapezoid is given by the formula $\frac{1}{2}$ (a+b) X h, where h is the height and 'a' and 'b' are the lengths of the two parallel sides as shown in the diagram below.

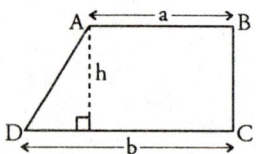

Example:

Find the area of a trapezoid with a =10 cm, b =12 cm and h =7 cm.

Solution: Area = $\frac{1}{2}$ (10+12) × 7

= 77 sq cm.

Area of a Triangle

The area of a triangle is equal to the product of one-half the base, b, times the height, h. The height (or altitude) of a triangle is a line drawn from a vertex perpendicular to the opposite side, called the base (see diagrams below).

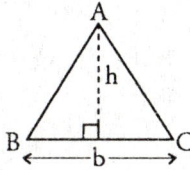 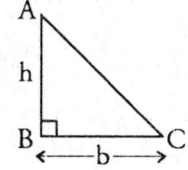 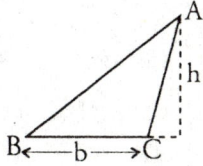

Example:

Find the area of the triangle below.

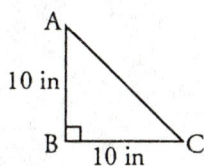

Solution: Area = $\frac{1}{2}$ b × h

But base = 10 and h =10

Therefore area = $\frac{1}{2}$ × 10 × 10 = 50 sq in

Area of a Circle

The area of a circle is given by $\frac{\pi d^2}{4}$ or πr^2 where d is the diameter, r is the radius and $\pi = \frac{22}{7}$.

Further Examples in Area Problems

1. Find the area of a rectangular field bounded at the ends by two semi circles as shown in the diagram below.

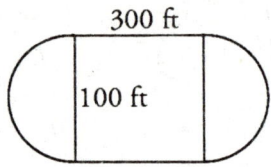

   Solution: The area of the figure is made up of the area of a rectangle plus the areas of two semi circles (or one circle).

   Thus area = $100 \times 300 + \frac{\pi d^2}{4}$ where d is the diameter, equal to 100 feet in the figure.
   Thus area = $100 \times 300 + 2500\pi$
   $= (30,000 + 2500\pi)$ sq ft.

2. Find the area of the triangle below.

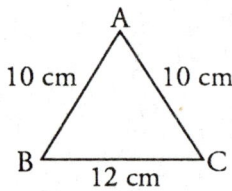

   Solution: Since this is an isosceles triangle, a line drawn from the vertex A to the base BC and perpendicular to it will bisect BC as shown opposite.

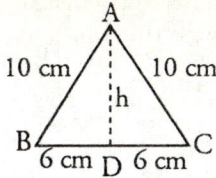

Using △ABD, we can calculate the height by using the Theorem of Pythagoras as follows:

$10^2 = 6^2 + h^2$
$h^2 = 10^2 - 6^2$
$= 100 - 36$
$= 64$
$h = 8$

With h = 8, and BC (base) = 12, the area of the triangle ABC is given by $\frac{1}{2}$ x 12 x 8 or 48 sq cm.

3. Find the area of the rectangle ABCD below.

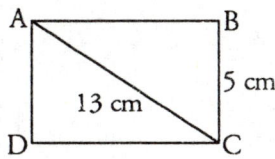

Solution: For the area we need to know the length of the figure. The width is given as 5 cm.

Using the Theorem of Pythagoras, the length AB is given by AB = $\sqrt{13^2 - 5^2}$
   = $\sqrt{144}$
   = 12

Thus the area = 12 x 5 = 60 sq cm.

4. If the shaded sector OAB of the circle below has an area of 10π sq in, what is the area of the whole circle?

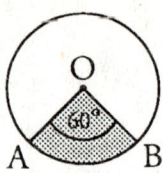

Solution: Since the angle of the sector is 60° and that for a full circle is 360°, it means that the sector is $\frac{60}{360}$ ($\frac{1}{6}$) of the full circle. Thus, the area of the full circle must be 6 times the area of the sector or 60π sq cm.

Practice Problems Involving Area

1. Find the area of (a) triangle OAB and (b) the shaded portion in the figure below. The radius of the circle is 6 cm.

2. The total area of the field shown below is 2,700 square yards. Find the area of the shaded triangle.

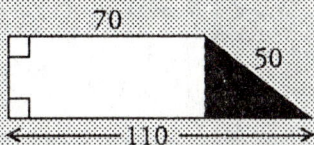

3. In the figure below, A and B are the centres of the two circles. If the area of the smaller circle is 24π, what is the area of the larger circle?

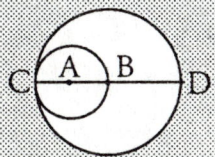

Answers: (1a) $9\sqrt{3}$ (1b) $6\pi - 9\sqrt{3}$ (2) 600 (3) 96π

3.5 Volumetric (Three-Dimensional) Figures

In a three-dimensional figure, the total surface of the solid is referred to as the *surface area*, and the total space contained within the figure is called the *volume*. Volume is expressed in cubic denominations. As in planar figures, all dimensions must be expressed in the same denomination before computing surface areas or volumes of solid figures.

Rectangular Solids

A *rectangular solid* (or block) is a figure having six rectangular faces that meet each other at right angles. The three dimensions of the figure are length (L), width (W) and height (H).

The surface area of the block is the sum of the surface areas of the six faces. This is given by LH + LW + HW + WL + HL + HW or 2 (LH + HW + LW).

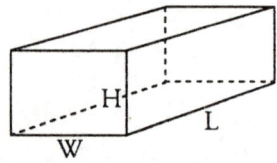

The volume of the block is given by the product of the length, width and height. Volume = L X W X H.

Example:

What is the volume of a rectangular block with length 10 cm, width 6 cm and height 5 cm?

Solution: Volume = 10 X 6 X 5
= 300 cu cm (cm^3)

A *cube* is a rectangular solid, the edges of which are equal.

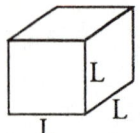

The surface area of a cube is 6 times the surface area of one face.

Example:

>Calculate the surface area of a cube of length 6 cm.
>*Solution:*   Total surface area = 6 (6 X 6) sq cm
>   = 216 sq cm.

Cylindrical Solids (Circular Cylinders)

A cylindrical solid is a three-dimensional figure with circular ends.

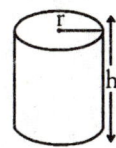

Referring to the figure above, the total surface area is the area of the curved surface plus twice the area of a circular face.
Thus surface area = h X $2\pi r$ + 2 X πr^2.

Example:

>The circular ends of a cylindrical solid have a radius of 5 cm. If the height of the solid is 20 cm, calculate its surface area.
>
>| | |
>|---|---|
>| *Solution:*   Area of two circular faces | = 2 X π X r^2 |
>| | = 2 X π X 25 |
>| | = 50π sq cm |
>| Area of the curved surface | = $2\pi r$ X h |
>| | = 10π X 20 |
>| | = 200π |
>| Thus the total surface area | = 250π sq cm. |

The volume of a cylindrical solid is given by the product of the cross-sectional area and the height v = πr^2 X h.

Example:

>If a circular cylinder has a radius of 7 in and a height of 10 in, then its volume equals π X 7^2 X 10 or 490π cu in.

Spheres

A sphere is a circular solid. The volume of a sphere is given by the formula $v = \dfrac{4\pi r^3}{3}$, where r is the radius of the sphere.

Practice Problems Involving Three-Dimensional Figures

1. If 1000 cubic feet of water is poured into a rectangular tank whose base has length and width of 10 and 6 feet respectively, how high does the water rise in the tank?

2. What is the volume of a sphere of radius 10 cm?

3. What is the total surface area of a cube of length 6 inches?

4. If a cylinder of radius 3 feet and length 6 feet is rolled on a level path so that it performs 10 revolutions, what area does it cover?

Answers: (1) 16.7 feet   (2) 1333.3π cm   (3) 216 square inches   (4) 360π square feet

3.6 Co-ordinate Geometry

Every point in a plane can be specified by measuring units (or co-ordinates) along two rectangular axes, usually called x and y. The co-ordinate axes are shown in the figure below.

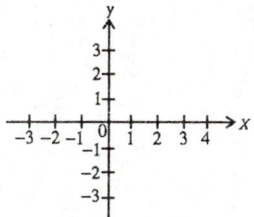

In such a system, every point can be specified by one x and one y co-ordinate, written as (x,y) and referred to as an ordered pair.

Example:

What are the co-ordinates for A in the figure below?

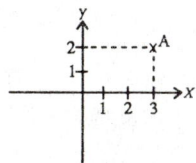

Solution: The co-ordinates of A are 3 for x and 2 for y or $(3,2)$.

The Equation of a Straight Line

The co-ordinate system can be used to graph any equation. Ordered pairs for an equation can be plotted thereby obtaining a graphical representation of the equation.

Example:

Graph $y = 2x$

Solution: A table of values (ordered pairs) is constructed by putting various values of x in the equation and calculating the corresponding y values.

x	−2	−1	0	1	2	3
y	−4	−2	0	2	4	6

These are then plotted and a straight line drawn through them. (Actually for a straight line only two ordered pairs need to be plotted and a straight line drawn through them.)

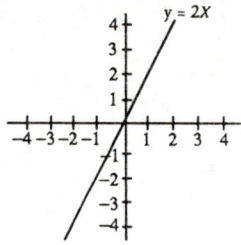

Conversely, a linear equation can be written for any straight line in a co-ordinate plane. The general form is given by

y = mx + c

where m is the slope (or gradient) of the straight line and c is its y–intercept (the point where the line cuts the y axis).

Example:

Write the equation of a straight line whose slope is –2 and which cuts the y axis at y = 3.

Solution: The equation is y = –2x + 3.

Also note that since only two points are required to specify a straight line, any two known points can be used to specify the equation of a straight line. If two points with co-ordinates (x_1, y_1) and (x_2, y_2) are known, then the equation of the line is deduced from the fact that the slope of a straight line is the same regardless of what portion of it is used to measure the slope. Thus if (x, y) are the co-ordinates of any point on the line, then in general,

$$\text{slope} = \frac{y_2 - y_1}{x_2 - x_1} = \frac{y - y_1}{x - x_1}$$

$$\text{or } y = \left(\frac{y_2 - y_1}{x_2 - x_1}\right)(x - x_1) + y_1$$

Graphing of Geometric Figures

The co-ordinate system may also be used for graphing geometric figures. The figure below is a rectangle whose vertices are A(0,4), B(6,4), C(6,0) and D(0,0).

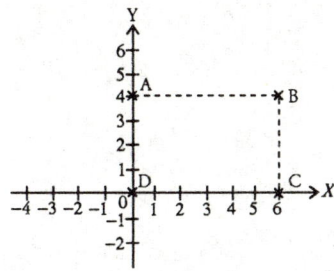

The lengths AB, BC, DC and AD can be obtained easily by subtraction as follows.

AB = 6–0 = 6 (subtracting the *x* variables)
BC = 4–0 = 4 (subtracting the y variables)

These lengths are easy to find since AB and BC are parallel to the *x* and y axis respectively. In the case where this is not so, such as when we are required to find the length DB say, we need to use the Theorem of Pythagoras.

The distance between two points $A(x_1, y_1)$ and $B(x_2, y_2)$

The distance between any two points in a co-ordinate plane is easily determined using the Theorem of Pythagoras. For any two points $A(x_1, y_1)$ and $B(x_2, y_2)$, the length AB is given by

$$AB = \sqrt{(x_2-x_1)^2 + (y_2-y_1)^2}$$

Example:

Find the length AB between points A (2,1) and B (5,4).

Solution: $AB = \sqrt{(5-2)^2 + (4-1)^2}$
 $= \sqrt{9+9} = \sqrt{18} = \sqrt{2 \times 9}$
 $= 3\sqrt{2}$

Practice Problems Involving Co-ordinate Geometry

1. Find the length of BD in the figure ABCD shown on page 49.
2. Find the area of the figure shown on page 49.
3. Find the area of the circle shown in the co-ordinate axes below.

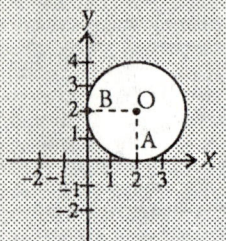

4. Find the distance between 2 points A and B with co-ordinates A (1,3) and B (–3, –5).

5. What is the area of a circle in a co-ordinate plane with centre P (10, 6), and a point Q on the circumference with co-ordinates (10, 2)?

Answers:

1. The length AB = 6 units (from above)
   The length AD = 4 units (from above)

   Thus $DB^2 = AB^2 + AD^2$
   $= 36 + 16$
   $DB = \sqrt{52}$
   $= \sqrt{4 \times 13}$
   $= \sqrt{4} \times \sqrt{13}$
   $= 2\sqrt{13}$

2. AB = 6 (above)
   AD = 4 (above)
   Area = 6 × 4 = 24 sq units

3. The circle is tangent to the *x* axis at A and to the *y* axis at B. The centre of the circle is the intersection of the perpendiculars at A(y=2) and B(x=2) which is (2,2). Thus the radius of the circle is 2 units and the area = πr^2 or 4π sq units.

4. 8.9

5. 16π

4. TABLES AND GRAPHS

4.1 Tables

Tables are used to organise information in a form that aids understanding. The title and margins (also called stubs) of the table are the key to the meaning of the information given in the table.

The table below represents the monthly sales in dollars of a small corner shop. Without the table and margins the individual entries would make no sense at all. It may even appear that some entries are sums of others, see the figure 1800 for March.

	Jan	Feb	March
1991	400	600	800
1992	300	700	1000
1993	600	1300	1800

It is important to take note of the units that are used in a table as these may vary from the units used in a problem based on the table.

Example:

The relationship between air pressure in lb/in² and the amount of rainfall in inches for Hypothetica is shown in the table below. If the pressure on a particular day is 200 lb/in², what is the expected level of rainfall?

Pressure (lb/in²)	Rainfall (in)
50	2.5
150	2.2
200	1.7
250	1.1
300	0.6

Solution: From the table a pressure of 200 lb/in² is associated with a rainfall of 1.7 in.

4.2 Graphs

Graphs are pictorial representations of data and information. They use distance or area to represent value and are used to illustrate comparisons and trends in statistical information. The most common graphs are line graphs, bar graphs, pie charts (circle graphs) and pictographs. Graphs are always labelled to indicate what each value means.

Line Graphs

Line graphs are used to show trends over a period of time and/or to compare trends for more than one variable. Examples are shown below.

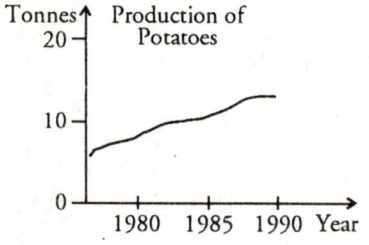 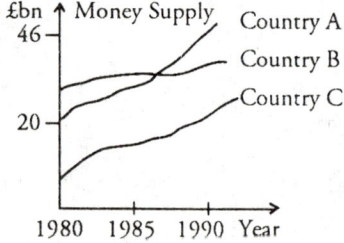

Example:

Which country, A, B, or C, has the fastest growth of money supply over the 1980-90 period?

Solution: The line graph with the steepest slope is that for Country A. Thus A has the highest growth of money supply.

Bar Graphs

Bar graphs are used to compare various quantities. Each bar may represent a single quantity or may be subdivided to represent several quantities (cumulative bars). Stacked bars may also be used to compare data from different sources over time. Examples of these are shown below.

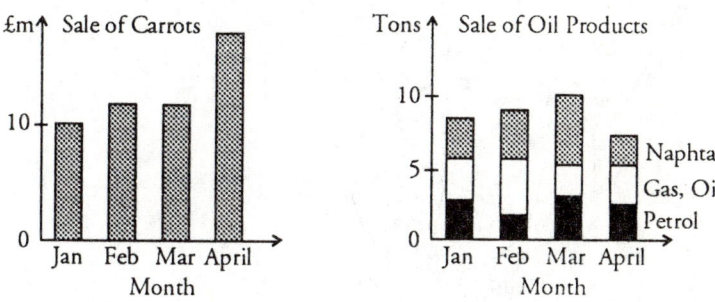

Example:

In the stack bar graph above, which country shows the fastest growth in its share of manufacturing in total GDP?

Solution: The share of manufacturing in Country C grew fastest in the 1980-90 period.

Pie Charts

Pie charts are used to show the proportions of various parts of a particular quantity. The proportions are usually expressed in percentage terms and the whole circle represents 100 per cent or 360°. Each part of the chart is called a section. The pie chart below shows the shares of government spending in a certain country, Dreamland.

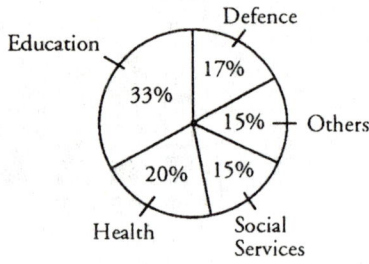

Example:

If the total government spending in Dreamland is $500m, how much is actually spent on defence?

Solution: Defence spending is 17% of the total which equals 0.17 × 500 m or $85 m.

Pictographs

Pictographs are used to compare quantities using symbols. Each symbol represents a given number of a particular item.
The sales of cars in a certain country in various years may be represented as follows:

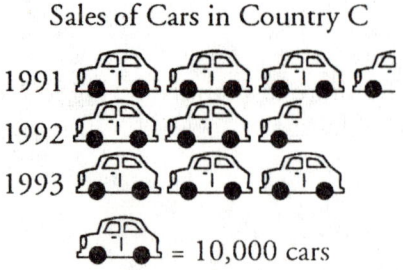

Examples:

1. In the pictograph above, how many cars were sold in Country C in 1992?

    Solution: Total number of cars sold was $2\frac{1}{2}$ × 10,000 or 25,000 cars.

2. How many fewer cars were sold in 1993 than in 1991?

    Solution: Total number of cars sold in 1993 equals 3 × 10,000 or 30,000.
    Total number of cars sold in 1991 was $3\frac{1}{2}$ × 10,000 or 35,000.
    Therefore 5,000 less cars were sold in 1993 than in 1991.

PART 2
How to Tackle GMAT Maths Questions

The proportion of maths questions in the GMAT has increased progressively over the last decade from about 30 to nearly 50 per cent. This reflects the increasing importance of mathematics in the analysis of business issues.

Two types of maths questions feature in the GMAT - Problem Solving and Data Sufficiency.

The standard breakdown of questions in the GMAT is as follows:

1. Problem Solving Section 1   16 questions   25 minutes
2. Problem Solving Section 2   16 questions   25 minutes
3. Data Sufficiency   20 questions   25 minutes

4. Reading Comprehension   18-23 questions   25-30 minutes
5. Critical Reasoning   16 questions   25 minutes
6. Sentence Correction   22 questions   25 minutes

7. AWA 1 - Analysis of an Issue   1 essay question   30 minutes
8. AWA 2 - Analysis of an Argument 1 essay question   30 minutes

9. A section containing experimental
   questions which is not marked   25 minutes

This section contains 48 examples of Problem Solving questions and 30 examples of Data Sufficiency questions. Solutions to these are provided and to make the best use of the example questions you should work through each problem before reading the solution.

Remember, the key to a high score on the GMAT lies in discovering and applying the quickest ways of solving each problem. It is important to read and fully understand the short cuts which are highlighted as these could make all the difference to your GMAT score.

Section 1

PROBLEM SOLVING

GMAT Problem Solving questions test (a) basic mathematical skills, (b) understanding of elementary mathematical concepts and (c) the ability to reason quantitatively and to solve quantitative problems.

Directions:
In this section solve each problem, using any available space on the page for rough work. Then indicate the best of the answer choices given.

Numbers:
All numbers used are real numbers.

Figures:
Figures that accompany problems in this section are intended to provide information useful in solving the problems. They are drawn as accurately as possible except when it is stated in a specific problem that its figure is not drawn to scale. All figures lie in a plane unless otherwise indicated.

Method of attack:
The steps required to arrive at the correct answer in a problem solving question are shown in the flow chart below. Two slightly different ways of arriving at the answer are indicated – working towards the answer or substituting the options in the problem to see which option fits a particular problem.

In conventional high school mathematics tests, marks are awarded for steps that lead to the final answer as well as the final answer. In the GMAT, it is only the answer that matters and any method that leads to the right answer in the shortest possible time is what is required. Speed and accuracy are both important. This section highlights the short cuts that are necessary for a high score on the test.

GMAT PROBLEM SOLVING
Flow Diagram Pattern of Attack

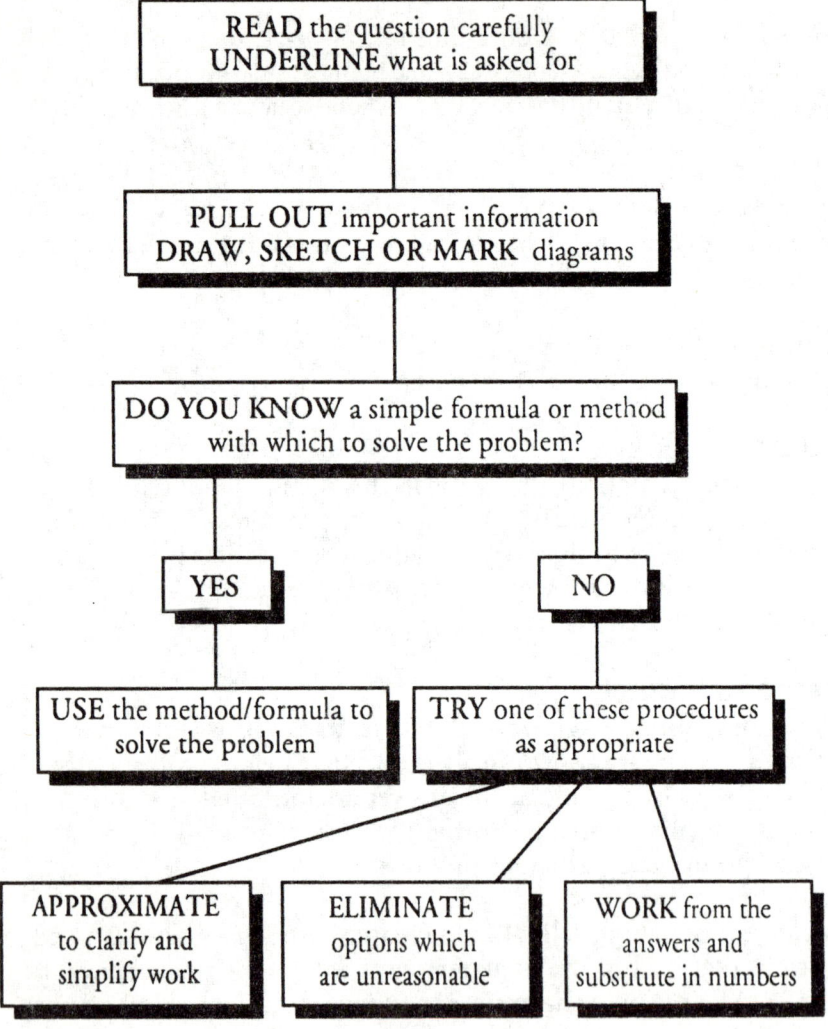

1. Evaluate $\dfrac{\frac{1}{4}\left(\frac{3}{2} \div \frac{4}{3}\right)}{\frac{2}{9}}$

   (A) $\dfrac{12}{8}$ (B) $\dfrac{81}{64}$ (C) $\dfrac{62}{81}$ (D) $\dfrac{78}{63}$ (E) $\dfrac{69}{32}$

Solution: Step 1 - simplify the top brackets: $\dfrac{1}{4}\left(\dfrac{3}{2} \times \dfrac{3}{4}\right)$ or $\dfrac{1}{4}\left(\dfrac{9}{8}\right)$

Step 2 - simplify the top brackets: $\dfrac{1}{4}\left(\dfrac{9}{8}\right) = \dfrac{9}{32}$

Step 3 - divide by denominator: $\dfrac{9}{32} \div \dfrac{2}{9}$

$$= \dfrac{9}{32} \times \dfrac{9}{2}$$

$$= \dfrac{81}{64}$$

The correct answer is (B).

2. In Figure 1, what is the value of $x + y$?

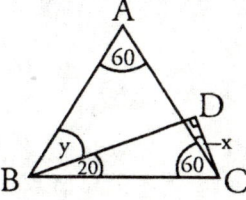

Figure 1

(A) 20   (B) 60   (C) 50   (D) 30   (E) 40

Solution: Note immediately that triangle ABC is equilateral and BCD is right angled.

Thus $\angle ABD + \angle DBC = 60°$.

Since $\angle DBC = 20°$, $\angle ABD$ (which is y) $= 40°$.

In $\triangle DBC$, $\angle BCD = 180° - 90° - 20° = 70°$.

But $\angle BCD = \angle BCA + \angle ACD$ and $\angle BCA = 60$

$\therefore \angle ACD$ (which is x) $= 10$.

Thus $x + y = 40 + 10 = 50$.

The correct answer is (C).

3. In a privatisation programme, John purchased y shares at a certain price. If in the first year after the purchase the price of a share increased by $0.50 and the total increase in John's shares was $75, how many shares did John purchase?

(A) 120   (B) 100   (C) 140   (D) 150   (E) 200

Solution: Let x = number of shares purchased; then the total increase in shares = $0.5x$.
But this is $75
$\therefore 0.5x = 75$
and $x = 150$.

The correct answer is (D).

Short cut: Notice that 100 shares will account for $50 of total increase in price and 200 will result in $100. The actual rise was $75 which is midway between $50 (100 shares) and $100 (200 shares). Thus the actual number must be 150, which is midway between 100 and 200 shares.

4. If 0.625K = 1.25 what is K?

   (A) 2.25   (B) 1.75   (C) 3.15   (D) 2.15   (E) 2

Solution: Conventionally the solution of this problem is as follows:
0.625K = 1.25.
Dividing both sides of the equation by 0.625, we have
$$K = \frac{1.25}{0.625} = 2$$
The correct answer is (E).

Short cut: GMAT requires candidates to be familiar with simple fractions which include eighths. Recognise that 0.625 is in fact $\frac{5}{8}$. Thus 0.625K = 1.25 is the same as $\frac{5}{8}$ K = 1.25 or K = 1.25 x $\frac{8}{5}$ or 2.

This avoids the problem of having to evaluate $\frac{1.25}{0.625}$ without a calculator which can be time consuming and difficult.

5. If $x > 0$ and $x^2 = 191$, what is the best whole number approximation of x?

   (A) 13 (B) 14 (C) 15 (D) 12 (E) 16

Solution: Every positive number has two roots, one of which is negative. The fact that $x > 0$ tells us that it is the positive root that is required.

Now $13^2 = 169$ and $14^2 = 196$, so the actual root is between 13 and 14 but 14^2 is closer to 191 than is 13^2. Therefore the whole number that is closest to $\sqrt{191}$ is 14.

The correct answer is (B).

6. At the Grand Studios, 5 tickets can be purchased together for the price of 4. What proportion of the price of 5 tickets is saved when purchased together?

(A) $\frac{1}{2}$ (B) $\frac{5}{4}$ (C) $\frac{1}{5}$ (D) $\frac{1}{4}$ (E) $\frac{4}{5}$

Solution: Let x = price of a ticket bought singly, then the price of 5 tickets = $5x$.
But 5 tickets bought together cost $4x$.
Thus the ratio of the price of 5 tickets bought together to that of 5 tickets bought in singles is $\frac{4x}{5x}$ or $\frac{4}{5}$

The price saved is thus $\frac{1}{5}$.

The correct answer is (C).

Short cut: 5 for the price of 4 simply means one pays for 4 instead of 5 or $\frac{4}{5}$, of what is due, implying a saving of $\frac{1}{5}$.

7. Andrea alone takes 6 hours to do a piece of work. Kathy alone takes $4\frac{1}{2}$ hours to do the same job. If Andrea worked on the job for 4 hours and leaves Kathy to finish the work, how long does it take Kathy to complete it?

(A) $2\frac{1}{2}$ (B) 2 (C) $1\frac{1}{4}$ (D) $1\frac{1}{2}$ (E) 3

Solution: If Andrea takes 6 hours to do the work independently then in 1 hour she does $\frac{1}{6}$ of the work and $\frac{4}{6}$ in 4 hours.

This leaves $\frac{2}{6}$ (or $\frac{1}{3}$) of the work for Kathy to complete.

For the whole work Kathy takes $4\frac{1}{2}$ hours.

For $\frac{1}{3}$ of the work she will take $\frac{1}{3}$ of $4\frac{1}{2}$ hours or $1\frac{1}{2}$ hours.

The correct answer is (D).

8. The rectangular tank shown in Figure 2 is filled with alcohol to a depth of 1.5 metres. If the tank is turned such that the tank rests on the face with sides 4 and 3 metres, how deep will the alcohol in the tank be?

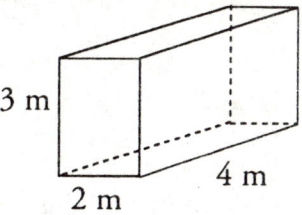

Figure 2

(A) 2   (B) 1.0   (C) 1.5   (D) 3   (E) 2.5

Solution: Volume of alcohol in the tank equals the product of the base area and the depth of alcohol.
That is 2 × 4 × 1.5 or 12 m³.
Let h = depth of the alcohol when the tank rests on the face with sides 3 and 4 metres.
Then volume of alcohol = 3 × 4 × h
but this is equal to 12 m³
∴ 12 h = 12 m
and h = 1.

The correct answer is (B).

Short cut: Notice that when the tank lies on the face with cross-sectional area 2 × 4 the depth of alcohol is 1.5 so when it lies on the face with a larger cross sectional area (4 × 3) the depth of alcohol will be less than 1.5. Looking at the options we can see that only (B) is less than 1.5. It is therefore not necessary at all to go through the above analysis.

9. A certain factory produces Z units of a product each working day. In one 25 working-day month, there was no production in the first y days as a result of a strike. How many units of the product should be produced per day in the remaining working days to still maintain an average of Z units per day?

(A) $\dfrac{25}{Z}$ (B) $\dfrac{25Z}{Z}$ (C) $\dfrac{25-y}{25-y}$ (D) $\dfrac{25Z}{25-y}$ (E) $25Z$

Solution: In 25 working days 25Z units should be produced.
Let number of strike days = y.
Thus number of actual days worked = 25−y.
Let the number of units required to be produced per day to maintain an average of Z units (or a total of 25Z units) = K.
Then K (25−y) = 25Z.

Thus K = $\dfrac{25Z}{25-y}$

The correct answer is (D).

10. The royalty on a GMAT book which sells for $20 is 10 per cent of annual sales with a guaranteed minimum of $1,000 each year. How much does the author receive in total in a 3 year period if 1000, 600 and 400 copies of the book were sold in the first, second and third years of publication?

(A) 4,200 (B) 4,000 (C) 3,800 (D) 5,000 (E) 4,600

Solution: The best way to solve this problem is to put the data in tabular form.

Yr	$ Sales	Royalty
1	1,000 × 20	2,000
2	600 × 20	1,200
3	400 × 20	800 + 200 (to reach minimum guaranteed)
Total		4,200

The correct answer is (A).

11. If x, y and z are 3 non-zero single digit integers and 100(x) + 50(y) + 20(z) = N, what is the units digit of N?

    (A) 2   (B) 1   (C) 0   (D) z   (E) y

Solution: All the coefficients of the variables are multiples of 10 and therefore N will have 0 as the last digit.
The correct answer is (C).

12. An extremely high demand for Wimbledon tickets in 1990 enabled a ticket tout to resell a $60 ticket for $240. What was the per cent increase in the price of the ticket?

    (A) 4   (B) 400   (C) 300   (D) 3   (E) 350

Solution: The tout got an increase of $240–60 or $180. $180 is 3 times $60 or 300 per cent of the price he paid.
The correct answer is (C).

Caution: Note that (A) would be right if the question had been 'how many times the cost was the resale price?'

13. If $10x - 3y = 8$, what is $6y - 20x$?

    (A) 16   (B) $3x - 16$ (C) $2y + 14$ (D) -14 (E) -16

Solution: Let us make 3y the subject of the equation.
That is $3y = 10x - 8$.
Subtracting 10x from both sides, we have $3y - 10x = -8$.
Comparing this with $6y - 20x$, we can see that the left hand side of the equation will be equal to this if we multiply the equation through by 2, that is

$2 \times (3y - 10x) = -8 \times 2$
or $(6y - 20x) = -16$.
Thus the value of $(6y - 20x)$ is -16.
The corrrect answer is (E).

Short cut: The left hand side of the equation will become $(-20x + 6y)$ if it is multiplied through by -2. The right hand side thus becomes -16 which is the answer we require.

14. 4 stacks containing equal numbers of coloured wooden blocks are to be made from 12 red blocks, 9 blue blocks and 7 green blocks. If all 28 blocks are used and each stack contains at least 1 of each colour, what is the maximum number of red blocks in each stack?

(A) 7   (B) 6   (C) 5   (D) 4   (E) 8

Solution: Each stack must contain 7 blocks since there are 28 blocks. If a stack has the minimum requirement of 1 each of blue and green blocks, the maximum number of red blocks in each stack must be 5. The first three blocks in each stack must be 1 red, 1 blue and 1 green, but there is no reason why the last four in a stack cannot be all red. The correct answer is (C).

15. A research officer's salary increased by 50 per cent in his second year of appointment. In his third year he was offered $2\frac{1}{2}$ times his first year salary. If the total gross salary earned in all 3 years is $90,000, what was his salary in the third year?

(A) $25,500   (B) $30,000   (C) $35,000
(D) $45,000   (E) $50,000

Solution: Let his first year salary be x.
Then, 2nd year salary: $x + 0.5x$.
3rd year salary: $2\frac{1}{2}x$.
Total for 3 years $= x + 1.5x + 2.5x$
but this is equal to $90,000.
Thus $5x = 90,000$ and $x = 18,000$.
The third year salary is $2.5x$ or $45,000.
The correct answer is (D).

Short cut: A first year salary of $30,000 is not possible because this is $90,000 divided by 3 and yearly salaries are not equal. So try using $20,000 as the figure for the first year. Then the second year salary must be $30,000, leaving $40,000 for the third year. But $40,000 is less than $2\frac{1}{2}$ times $20,000. That is the $20,000 assumed for the first year is slightly high. A figure less than $20,000 will leave a higher third year salary. The third year salary should be more than $40,000 and the first year less than $20,000. Looking at the options it is only (D) which is higher than $40,000.

16. How many 2 digit whole numbers yield a remainder of 1 when divided by 10 and also yield a remainder of 1 when divided by 8?

    (A) 1   (B) 2   (C) 3   (D) 4   (E) 5

Solution: The only 2 digit numbers that leave a remainder of 1 when divided by 10 are 11, 21, 31, 41, 51, 61, 71, 81, 91. Of these only 41 and 81 are divisible by 8 with a remainder of 1.

The correct answer is (B).

17. If 3 identical rectangles form a square when placed together as shown in Figure 3 and the length of each individual rectangle is x times its width, then x is

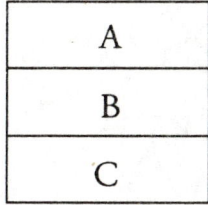

Figure 3

(A) 1 (B) 2 (C) 3 (D) 4 (E) 5

Solution: If the figure formed is a square then the width and the length must be equal. But the overall width is formed from 3 times the width of each individual rectangle. Put differently the length is three times the width of the individual rectangles.

The correct answer is (C).

18. What is the area in square feet of the playing field with semi-circular ends shown in Figure 4.

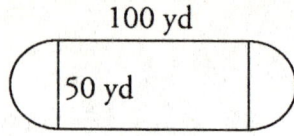

Figure 4

(A) 4,000 + 900π (B) 5,000 + 625π (C) 5,000 + 312.5π
(D) 45,000 + 5625π (E) 45,000 + 2812.5π

Solution: We have to find the area of the rectangular part and the two semi-circular parts and add them together. Remember that the answer must be given in square feet. It is much easier to convert the figures into feet at the onset.

Area of the rectangular part = 50 X 3 ft X 100 X 3 ft
= 45,000 ft².
Area of 2 semi-circles = area of one circle

$$= \pi \times \left(\frac{150}{2}\right)^2$$

$$\left(NB\ \pi\ \frac{d^2}{4}\right)$$

$$= 5625\pi$$

The total area = 45,000 + 5625π. The correct answer is (D).

Caution: Note that if the area is calculated in square yards the answer will be option (B). If you make the mistake of not reading the question and miss out the fact that the answer must be in square feet, you will obtain the answer in option (B) and choose the wrong option.

Short cut: You can close in on the right answer by first finding the area of the rectangular part i.e. 150 ft X 300 ft = 45,000. With this, only options (D) and (E) can be correct. Then you can home in on the right choice by noting that the radius of the semi-circle is $\frac{150}{2}$ ft (75). The area of a circle is πr² and the square of 75 will have 5 at the end. Of (D) and (E), (D) must be right.

19. If $\dfrac{Z^2 - 25}{3x} = Z + 5$ for $Z \neq -5$, then in terms of x, $Z =$

   (A) $3x + 5$ (B) $3x^2 + 5$ (C) $x + 5$ (D) $5 + x$ (E) $5x + 3$

Solution: Factorise the numerator, obtaining

$$\dfrac{(Z + 5)(Z - 5)}{3x} = Z + 5$$

Since $Z \neq -5$, it implies that $(Z + 5) \neq 0$.
As a result we can divide both sides of the equation by $(Z + 5)$, leaving
$\dfrac{Z - 5}{3x} = 1$, which simplifies
to $Z = 3x + 5$. The correct answer is (A).

20. A trip to Calgary involves flying for x miles and then driving for y miles. If it costs 15 cents per mile to fly and 12 cents per mile to drive what is the cost of the trip in dollars?

   (A) $12x + 15y$ (B) $\dfrac{12x + 15y}{100}$ (C) $\dfrac{5x + 12y}{100}$

   (D) $\dfrac{15 + 12}{100}$ (E) $\dfrac{15x + 12y}{100}$

Solution: Here too we want to avoid the trap discussed in Q18. The answer must be in dollars although the figures given are in cents.

   Distance covered by air = x miles
   Cost per mile = 15 cents
   Total cost for flying = $15x$ cents
   Similarly cost for driving = $12y$ cents
   Total cost for the entire journey = $\dfrac{15x + 12y}{100}$

The correct answer is (E).

Caution: Note that (A) is right but in cents rather than dollars.

21. Which of the following numbers is equal to 82 per cent of 170?

   (A) 82.4   (B) 78.2   (C) 112.4   (D) 172.0   (E) 139.4

Solution: The answer to this is the product of 0.82 × 170 which equals 139.4. The correct answer is (E).

Short cut: This multiplication may take a couple of minutes. On a GMAT every second saved helps so let us see how we can get to the answer without really doing the multiplication. Note that $\frac{1}{2}$ (50 per cent) of 170 is 85. Since 82 per cent is bigger than 50 per cent options (A) and (B) are out of the question and they can be eliminated. Option (D) is also out because it is greater than 170 leaving (C) and (E). Now $\frac{3}{4}$ of 160 is 120 so $\frac{3}{4}$ (or 75%) of 170 is greater than 120 let alone 82%, thus option (C) is also out leaving (E) as the right answer.

22. If the square ABCD is rotated on a plane anticlockwise about its centre such that A moves into the position originally occupied by D, by how many degrees will the square ABCD have rotated?

   (A) 300   (B) 210   (C) 180   (D) 270   (E) 290

Solution: The easiest way to see what is happening in this problem is to draw lines AC and DB as shown here.

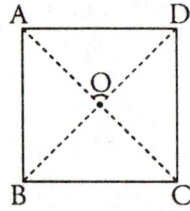

Since ABCD is a square AC and DB will pass through the centre O and they will be perpendicular to each other.
Thus the rotation from A to B is 90°, to C another 90°, and to D another 90° or 270° in all.
The correct answer is (D).

23. The overtime wage in a certain firm is 1.5 times the regular hourly wage, the latter being $8.40. If an employee works 8 hours overtime how much money does he get?

(A) $126.00 (B) $67.20 (C) $100.80 (D) $84.00
(E) $12.60

Solution: The overtime hourly wage   = 1.5 × $8.40
                                            = $12.60
For 8 hours the total money earned   = 8 × $12.60
                                            = $100.80

The correct answer is (C).

24. In 1989, a retailer received 12.5 per cent of the proceeds from the sale of a car. If the retailer again received 12.5 per cent of the proceeds of a similar car in 1990 but earned $200 more than he got in 1989, how much did the car sell for in 1990 if the 1989 price was $8,000?

(A) $9,200 (B) $9,600 (C) $8,000 (D) $8,200
(E) $9,000

Solution: In 1989 the retailer got 12.5 per cent of $8,000. Noting that 12.5 per cent is the same as $\frac{1}{8}$, the retailer got $\frac{1}{8}$ of $8,000, which is $1,000. In 1990, he got another 12.5 per cent but the actual amount was $200 more than in 1989, that is $1,200. If this is 12.5 per cent of the price of the car then the price must be 8 × 1,200, that is $9,600.

The correct answer is (B).

25. Between which of the following numbers is $\sqrt[4]{396}$?

(A) 1 and 2 (B) 2 and 3 (C) 3 and 4 (D) 4 and 5
(E) 5 and 6

Solution: The fourth root of a number is the square root of the square root.
Thus $\sqrt[4]{396} = \sqrt{\sqrt{396}}$ which is 4.46.
The correct answer is (D).

Short cut: If you recognise that 396 is very nearly 400, then you can solve the problem easily. Take the square root of 400 twice $\sqrt{400} = 20$ and 20 is greater than $4(\sqrt{16})$ and less than $5(\sqrt{25})$, so the fourth root of 396 is between 4 and 5, which is (D).

26. In 1980, 32 per cent of the proceeds from the sale of a product accrued to the producer. Five years later in 1985 only 24 per cent of proceeds accrued to him. If the price of the product was $67 in 1980 and $102 in 1985, which of the following is closest to the increase in the amount received by the producer from 1980 to 1985?

(A) $7   (B) $5   (C) $3   (D) $9   (E) $10

Solution: Attempts to actually compute 32 per cent of $67 and 24 per cent of $102 will be time consuming and difficult. But if we recognise that 32 per cent is very nearly equal to $\frac{1}{3}$ (33.3 per cent) and 24 per cent is very nearly equal to $\frac{1}{4}$ (25 per cent) and 102 is to all intents and purposes 100, then we have a neat straightforward solution.
Thus $\frac{1}{3}$ of 67 = 22 (approximately).
$\frac{1}{4}$ of 100 = 25 (approximately).
Difference = 25 − 22 = $3
The correct answer is (C).

Compare the results obtained by actual multiplication
32 per cent of 67 = 0.32 × 67 = 21.4
24 per cent of 102 = 0.24 × 102 = 24.5
Difference = 3.1, which is approximately 3.

74

27. If $2y = 5z$, $3x = 7y$ and $z \neq 0$, what is the ratio of x to z?

(A) $\dfrac{35}{6}$ (B) $\dfrac{6}{35}$ (C) $\dfrac{14}{15}$ (D) $\dfrac{21}{10}$ (E) $\dfrac{10}{21}$

Solution: We are required to write an equation that gives the value of $\dfrac{x}{z}$.

From the first equation, $z = \dfrac{2}{5}y$

and from the second $x = \dfrac{7}{3}y$.

Thus $\dfrac{x}{z} = \dfrac{\frac{7}{3}y}{\frac{2}{5}y} = \dfrac{7}{3} \times \dfrac{5}{2} = \dfrac{35}{6}$

The correct answer is (A).

28. A computer merchant bought some floppy diskettes for $120 which he later sold at $10 for 6 units. If he made a profit of $80 on the sale of the entire lot, how many diskettes did he buy?

(A) 90   (B) 100   (C) 110   (D) 120   (E) 130

Solution: Let the number of diskettes = x.
The price paid for the lot = $120.
He sold each unit for $\dfrac{\$10}{6}$.

The total sum obtained from the lot = $\dfrac{\$10}{6} x$.

Profit = total receipts − total costs, or

$\dfrac{10x}{6}$ − 120. But this equals $80.

Thus $80 = \dfrac{\$10}{6} x - 120$ and $x = 120$ diskettes.

The correct answer is (D).

29. 50 per cent of the total workforce of 1,000 employees in an institution are academic staff and 52 per cent are female. If 30 per cent of all female employees are academic staff, how many male employees are academic staff?

(A) 500   (B) 344   (C) 260   (D) 300   (E) 376

Solution: Since 52 per cent of the entire workforce of 1,000 are females, it means that 520 are females and 480 are males.
30 per cent of all female staff are academic
thus female academic staff = 0.3 × 520 = 156.
Total number of academic staff (50%) = 500.
Number of male academic staff = total − female = 500 −156 = 344.
Thus total number of male academic staff is 344.

The correct answer is (B).

30. The scoring system on the GMAT is such that one-quarter of the number of wrongly answered questions is subtracted from the total of correctly answered questions. If on a 100-question GMAT, a candidate answered all the questions and scored 70 marks, how many questions did the student answer incorrectly?

(A) 20   (B) 24   (C) 28   (D) 32   (E) 36

Solution: No. of questions answered = 100.
Let number of questions answered incorrectly = w.
Then number of questions answered correctly = 100 − w.
From this deduct one-quarter of questions answered incorrectly, that is $(100 - w) - \frac{1}{4}w$

but this equals 70.
Thus $(100 - w) - \frac{1}{4} w = 70$, giving w = 24.

The correct answer is (B).

31. If O is the centre of the circle shown in Figure 5 and OA = AB = BC, what is the value of y?

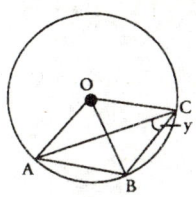

Figure 5

(A) 10   (B) 20   (C) 30   (D) 40   (E) 50

Solution: Since OA = AB, △OAB is equilateral
and ∠OAB = ∠ABO = ∠BOA = 60°.
Also since BC = AB = OA = radius of the circle, OBC is also equilateral and ∠OBC = ∠BCO = ∠COB = 60°
Now consider ABC. Since AB = BC this triangle is isosceles and ∠CAB = ∠ACB = y.
But ABC = 60 + 60 = 120°
thus 2y = 180–120 and y = 30.

The correct answer is (C).

32. There are between 119 and 138 tennis balls in a basket. If they are counted out 6 at a time or 7 at a time there are 4 left over. How many balls are in the basket?

(A) 119   (B) 120   (C) 126   (D) 130   (E) 138

Solution: The set of numbers between 119 and 138 divisible by 6 are: 120, 126, 132, 138 and those divisible by 7 are: 119, 126, 133.
126 is divisible by both 6 and 7. Thus 126 + 4 will be divisible by 6 or 7 with a remainder of 4.
The correct answer is (D).

33. Given that the ratio of the areas of two squares is 3:1, what is the ratio of the perimeters of the two squares?

(A) 1:3   (B) 2:3   (C) $\sqrt{2}:\sqrt{3}$ (D) $\sqrt{3}:1$ (E) $\sqrt{3}:2$

Solution: Let the area of the small square be x, then the area of the large one will be $3x$.

Area = length × breadth = (length)^2 for a square.

Thus the length of the bigger square = $\sqrt{3x}$ and the length of the smaller one = $\sqrt{x}$.

The perimeter of the bigger square is $4 \times \sqrt{3x}$ and the perimeter of smaller square is $4\sqrt{x}$.

Thus the ratio of the perimeters is $4\sqrt{3x} : 4\sqrt{x}$, which equals $\sqrt{3}:1$.

The correct answer is (D).

34. A company has 5,000,000 shares. Before the stock market crash of 1987, a share was worth 140 cents. If after the crash the company's share price fell by 10 per cent, how much was the company worth after the crash?

(A) $6.0m   (B) $6.3m   (C) $6.5m   (D) $6.7m
(E) $7m

Solution: After the crash the company's share price fell by 10 per cent to 126 cents per share. The original worth of the company was $1.4 × 5m or $7 million.

After the crash this declined by 10% or $\frac{1}{10} \times 1.40 \times 5m$ = $700,000.

Thus the company was worth $7m − $0.7m or $6.3m.

The correct answer is (B).

35. The mathematical operation ⊗ is defined in such a way that a ⊗ b = (a–b) + ab. For what value of a is a ⊗ b equal to b for all values of b?

(A) $b+1$ (B) $\dfrac{a-b}{ab}$ (C) $\dfrac{1+b}{2}$ (D) $\dfrac{b}{2+b}$ (E) $\dfrac{2b}{1+b}$

Solution: a ⊗ b = (a – b) + ab
but a ⊗ b = b (given).
Thus (a–b) + ab = b
factorising a(1+b) = 2b
and a = $\dfrac{2b}{1+b}$

The correct answer is (E).

36. A ladder of length 26 feet rests against a vertical wall. The bottom of the ladder is 10 feet from the base of the wall. If the top of the ladder slips down 5 feet, approximately how many feet will the bottom of the ladder slide?

(A) 5.1   (B) 7.7   (C) 9.2   (D) 10.0   (E) 11.0

Solution: By the Theorem of Pythagoras the height (h) of the ladder is $\sqrt{26^2 - 10^2}$ which is 24 ft.

If the top slides down 5 feet the height will drop to 19 ft.

The length of the ladder does not change, thus the base of the ladder will be $\sqrt{26^2 - 19^2}$ or 17.7 ft from the base of the wall, this representing a shift of 7.7 ft from the original position.

The correct answer is (B).

Short cut: The first point we have to recognise here is that the root to the solution of this problem is the Theorem of Pythagoras. The next step is to draw a diagram to represent the two situations followed by the use of the theorem to evaluate distances. No doubt without a calculator terms like $\sqrt{26^2 - 10^2}$ and $\sqrt{26^2 - 19^2}$ can be difficult and time consuming. But if we realise the basic Pythagoran relationships – 5, 4, 3 (ie $5^2 = 4^2 + 3^2$) and 13, 12, 5 (ie $13^2 = 12^2 + 5^2$) and use approximations, this problem need not cause any worry.

First the height of the ladder in situation 1 is given by $26^2 = h^2 + 10^2$ (1). If we realise that $13^2 = 12^2 + 5^2$, we can easily deduce that h in equation (1) must be 24 because the numbers in (1) are double those in $13^2 = 12^2 + 5^2$.

Thus we obtain the height of the top of the ladder. If the top drops 5 ft then the height falls to 19 ft. Thus the distance of the bottom of the ladder from the wall will increase to $\sqrt{26^2 - 19^2}$. We know that 26^2 equals 676. 19^2 can be approximated to 20^2 which is 400. Thus $\sqrt{26^2 - 19^2}$ is approximately equal to $\sqrt{676 - 400}$ which equals $\sqrt{276}$. 17^2 is 289 so $\sqrt{276}$ will be close to 17. In other words the ladder slides about 7 ft on the ground.

Looking at the answers the closest is (B).

37. A box contains 100 tennis balls, half of which are size 1 and the other half size 2. 25 per cent of the balls are red and the remaining 75 per cent green. If 20 size 1 balls are red, how many are size 2 and green?

   (A) 25   (B) 50   (C) 5   (D) 30   (E) 45

Solution: Making a table is the easiest way to tackle this problem just as drawing a diagram facilitates the solution of problems, such as the one in Q16.

	Total balls	Size 1	Size 2
Red	25	20	* (5)
Green	75	* (30)	*(45)
	100	50	50

We need to fill in the * marks to reach the answer. Since 20 size 1 balls are red and there are 25 red balls it implies that 5 red balls are size 2. Since there are 50 of size 1 and 50 of size 2, we deduce that there are 30 green size 1 and 45 green size 2 balls (see brackets). The correct answer is (E).

38. If $2x + 5y = 16$ and $6x - 4y = 0$, what is the value of $(2x + y)$?

   (A) $\frac{11}{19}$ (B) $\frac{112}{9}$ (C) $\frac{98}{19}$ (D) $\frac{112}{19}$ (E) 112

Solution: From the second equation: $y = \frac{6x}{4}$ substituting this into the first equation, we have $2x = 16 - 5\frac{6x}{4}$. Thus $x = \frac{32}{19}$. Putting this back into $y = \frac{6x}{4}$ we have $y = \frac{48}{19}$. Thus $2x + y = 2\frac{32}{19} + \frac{48}{19} = \frac{112}{19}$. The correct answer is (D).

Short cut: Since from the equation given, $y = 1.5x$, substituting this into the first equation we obtain $9\frac{1}{2}x = 16$ which is less than 2. If we approximate x to 2, then $y = 3$, so $2x + y = 7$. That is, the answer must be close to but less than 7. Looking at the options we can eliminate options (A), (B), (C) and (E), leaving only (D).

39. What is the least possible product of 4 different integers each of which has a value between −6 and 5 inclusive?

(A) −240 (B) −300 (C) −360 (D) −600 (E) −420

Solution: The least product is the largest negative number. The product of 4 numbers is negative if one of them is negative and the rest positive or 3 of them are negative and one is positive. The largest absolute numbers in the given set are 6, 5, 4, and 3. Thus the largest negative number is −6 X −5 X −4 X 5 = −600.

The correct answer is (D).

40. A traveller is required to obey speed limits on his journey from A to C via B. Travelling at 60 mph he covers the distance between A and B in 1.5 hours. It takes another 2.5 hours to cover the distance from B to C. If his car does 30 miles per gallon when travelling at 60 mph and 20 per cent less when travelling at 50 mph, what is the total amount of petrol consumed when the traveller covers the journey from A to C?

(A) 8.2 (B) 9.2 (C) 7.0 (D) 10.4 (E) 6

Solution:
The distance between A and B = speed X time
= 1.5 X 60
= 90 miles

The distance between B and C = 2.5 X 50
= 125 miles

Petrol consumed between A and B = $\frac{90 \text{ miles}}{30 \text{ mpg}}$
= 3 gallons

Petrol consumption between B and C = 30 less 20% or 24 mpg

Petrol consumed between B and C = $\frac{125}{24}$
= 5.2 gallons

Thus total petrol consumed is 8.2 gallons.

The correct answer is (A).

Short cut: Total journey = 60 × 1.5 + 50 × 2.5 = 215 miles.
If whole journey is covered at 60 mph, petrol required will be $\frac{215}{30} = 7.2$ gallons.

If journey is covered at 50 mph, petrol consumed will be $\frac{215}{24}$ or about 9 gallons.
That is, the answer is between 7.2 and 9 which is option (A).

41. A certain bank has 437 branches and 1 headquarters. If each branch has 11 unionised workers and the headquarters has 17, how many unionised workers are employed by the bank?

    (A) 4017 (B) 3748 (C) 4371   (D) 4824   (E) 5428

Solution: The number of unionised workers is 11 × 437 + 17
= 4824.

The correct answer is (D).

Short cut: Multiplying 11 by 437 and then adding 17 to the result, without a calculator will require over a minute, even for the smart student.
You can save time by using the method of approximation.
Since 437 × 10 equals 4370, the right answer will be greater than 4370. Options (A), (B) and (C) can be eliminated straight away, leaving (D) and (E). The right answer is 4370 plus 437 plus 17 which is less than 5000.

42. Electricity consumption in each of the 3 spring months is double that of each summer month and one-half that of each of the autumn and winter months. If the total consumption of electricity in a year is $1,650, how much is consumed in each summer month?

(A) $100   (B) $75   (C) $50   (D) $40   (E) $30

Solution: Let the consumption in each summer month be $x.
Then the consumption in each spring month is $2x.
The consumption in each autumn month is $4x.
The consumption in each winter month is $4x.

Total consumption in the year is

$3(x + 2x + 4x + 4x)$
but this is equal to $1650.
Thus $3 \times 11x = 1650$.

$$x = \frac{1650}{33} = \$50$$

The correct answer is (C).

43. A certain supermarket gives a 10 per cent discount when a telephone order is made and a further 10 per cent at the time of collection of the goods. What is the supermarket's percentage purchase discount?

(A) 10   (B) 15   (C) 19   (D) 20   (E) 25

Solution: Let the price of the telephone be p. Then, on order, a 10% discount makes the price go down to 0.9p
($p - 0.1p = 0.9p$)
On actual purchase the price 0.9p goes down further by 10%, giving a figure of 0.9 (0.9p) or 0.81p. Thus the total discount is 0.19 or 19%.

The correct answer is (C).

44. 40 per cent of replanted nursery plants were pear trees, the remainder being apple trees. If some of them died and 30 per cent of those that died were pears, what was the ratio of the death rate among pears to the death rate among apples?

(A) $\frac{3}{7}$ (B) $\frac{4}{6}$ (C) $\frac{2}{7}$ (D) $\frac{9}{14}$ (E) $\frac{3}{2}$

Solution: Let the total number of replanted plants be x.
Let the total number of plants which died be y.
Then the number of replanted pears = $0.4x$
and the number of replanted apples = $0.6x$.
Number of replanted pears which died = $0.3y$.
and number of replanted apples which died = $0.7y$.

Death rate among pears = $\frac{0.3y}{0.4x}$ (1)

Death rate among apples = $\frac{0.7y}{0.6x}$ (2)

The ratio of (1) to (2) = $\frac{\frac{0.3y}{0.4x}}{\frac{0.7y}{0.6x}}$

= $\frac{9}{14}$ The correct answer is (D).

45. The hire purchase arrangement in a stereo shop consists of a down payment of 10 per cent of the purchase price and 12 equal monthly instalment payments made up of the remainder of the purchase price plus 10% interest. If a man buys a stereo for $120 on the plan, what is his monthly payment?

(A) $12.00   (B) $11.00   (C) $9.90   (D) $9.00   (E) $8.50

Solution: Purchase price of stereo = $120.
Down payment is 10% of 120 = $12 leaving $108 to pay.
Add 10% of $108 = $119.
This is then divided by 12 yielding a monthly payment of $9.90.

The correct answer is (C).

46. The population of a certain country in 1969 was 99 m. By 1990 this had increased to 161 m. What was the per cent increase approximately?

(A) 50% (B) 63% (C) 69% (D) 70% (E) 73%

Solution: The percent increase in the population is given by the formula $\dfrac{\text{new population} - \text{initial population}}{\text{initial population}} \times 100$

$= \dfrac{(161 - 99)}{99} \times 100$

$= \dfrac{62}{99} \times 100 = 62.6\%$ or 63% (approx).

The correct answer is (B).

Short Cut: In evaluating $\dfrac{62}{99} \times 100$, note that 99 is very nearly 100, thus the answer must be close to 62%, which is option (B).

47. For a positive integer n, the number n! (called factorial n) is defined as n (n–1) (n–2) ... (1) for example 3! = 3 X 2 X 1. What is 6! – 4!?

(A) 620 (B) 720 (C) 596 (D) 696 (E) 540

Solution: 6! = 6 X 5 X 4 X 3 X 2 X 1
4! = 4 X 3 X 2 X 1

Thus 6! – 4! = 6 X 5 X 4 X 3 X 2 X 1 – 4 X 3 X 2 X 1
= 720 – 24
= 696

The correct answer is (D).

48. In how many ways can you arrange the letters in the word COURSE?

   (A) 30   (B) 120   (C) 240   (D) 720   (E) 960

Solution: This is a straightforward permutation problem. The number of letters in COURSE is 6. Thus the answer is 6! or 6 X 5 X 4 X 3 X 2 X 1 which equals 720.

The correct answer is (D).

Section 2

DATA SUFFICIENCY

GMAT Data Sufficiency questions are designed to test the candidate's ability (a) to analyse a quantitative problem, (b) to recognise the relevance of a piece of information and (c) to determine at what point there is enough information to solve a particular problem. Each Data Sufficiency problem consists of a question, often accompanied by some initial information, and two statements labelled (1) and (2), containing additional information. The candidate is required to decide whether sufficient information to answer the question is provided by either (1) or (2) individually, or if not, by both combined.

Directions

Each of the data sufficiency problems below consists of a question and two statements labelled (1) and (2), which contain some information. You have to decide whether the data or information given in the two statements are sufficient for answering the question. Using the data given in the statements plus your knowledge of mathematics and everyday facts (such as number of days in April or the meaning of terms such as clockwise), you choose option
(A) if statement (1) ALONE is sufficient, but statement (2) ALONE is not sufficient to answer the question asked;
(B) if statement (2) ALONE is sufficient, but statement (1) ALONE is not sufficient to answer the question asked;
(C) if both statements (1) and (2) TOGETHER are sufficient to answer the question asked, but NEITHER statement ALONE is sufficient;
(D) if EACH statement alone is sufficient to answer the question asked; and
(E) if statements (1) and (2) together are not sufficient to answer the question asked and additional data specific to the problem is needed.

Numbers

All numbers used are real numbers. The GMAT avoids the use of complex numbers in whatever form.

Figures

A figure in a data sufficiency problem will conform to the information given in the question, but will not necessarily conform to the additional information given in statements (1) and (2).

You may assume that lines shown as straight are straight and that angle measures are greater than zero.

You may assume that the positions of points, angles, regions etc. exist in the order shown.

All figures lie in a plane unless otherwise indicated.

Method of attack

An important point to note here is that you are not required to solve a problem; you need only to determine whether sufficient information is given to solve the problem. Read the question very carefully and then consider statement (1). The next stage is to consider statement (2).

Be sure to disregard all the information in statement (1) when considering statement (2). It is important that statements are considered independently of each other as failure to do so could result in the choice of a wrong option. The process for dealing with this section is summarised in the flow chart on the following page.

GMAT DATA SUFFICIENCY

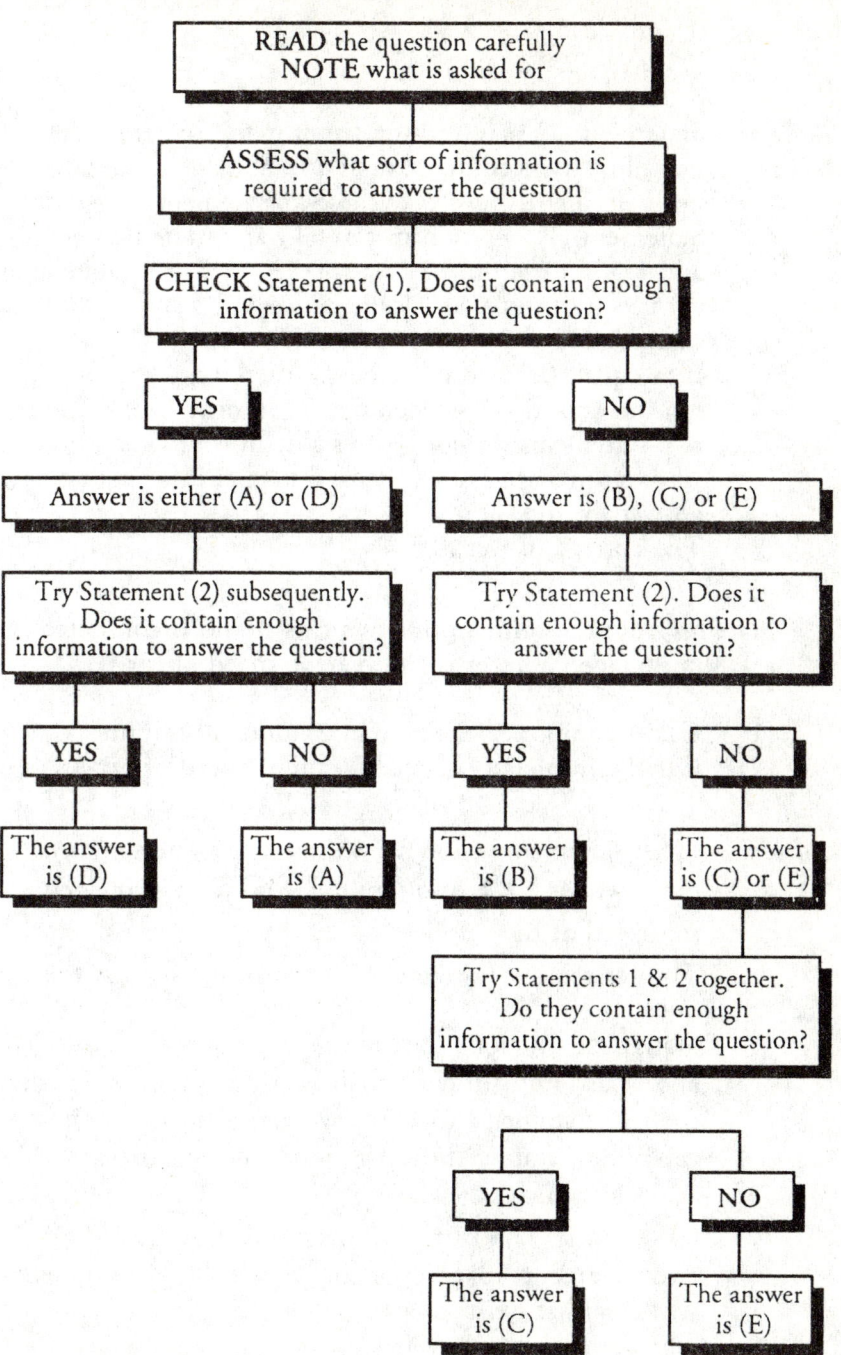

1. Is m > n?

    (1) 0 < m < 1.5
    (2) 0.5 < n < 2.0

Solution: Statement (1) says nothing about n and therefore does not allow us to compare m and n. Statement (2) does not help. Let us now take the two statements together. Statement (1) says that m can take any value between 0 and 1.5, for example 0.2, 0.5, 0.9, 1, 1.2, 1.4. Statement (2) says n can take any value between 0.5 and 2, for example 0.6, 0.9, 1, 1.4, 1.8, 1.9.
If m equals 0.2 and n = 0.6, say, then n > m but if n = 0.6 and m = 1.4, then m > n. In other words these two statements do not give us a definite answer as to whether m > n or not. We need further information to answer the question.
The correct answer is (E).

2. Starting from the same point, two cars A and B competed in a 100 mile race. What was the average speed of car B?

    (1) Car A completed the race in 2 hours 30 minutes.
    (2) Car B completed the race 25 miles ahead of car A.

Solution: From statement (1) above we can determine the speed of car A which is $\frac{100}{2.5}$ or 40 mph but it says nothing about the speed of B.

Thus statement (1) alone is not sufficient to answer the question.
Statement (2) alone does not give us the speed of car B and is also not sufficient to answer the question. Putting the two statements together, we note that by the time car B has completed the 100 mile race, A has covered 75 miles which takes it $\frac{75}{40}$ or $\frac{15}{8}$ hours. This implies that car B takes $\frac{15}{8}$ hours to complete the race. Thus the two statements together enable us to solve the problem while neither on its own is sufficient.
The correct answer is (C).

3. The pie-chart shown in Figure 1 indicates steel consumption by sector in country REF. How much steel was consumed by the automotive sector in 1990?

    (1) Electrical engineering consumed $10m of steel in 1990.
    (2) Household goods required $17m of steel in 1990.

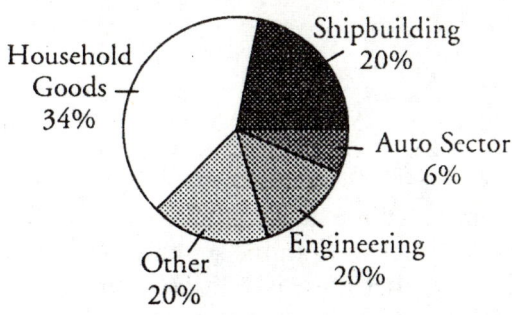

Figure 1
Shares of Steel Consumption in REF, 1990

Solution: The key to answering the question is the total consumption of steel in 1990. Once this is known it is a simple matter of computing 6 per cent of that figure which represents the level of steel consumption by the automotive sector.

Statement (1) says that the level of consumption of steel by the electrical engineering sector was $10m. The pie chart says that this represents 20 per cent of the total. Thus the total can easily be computed as $\frac{\$10}{0.2}$ or $50m. This enables the consumption by the automotive sector to be computed, that is $\frac{6}{100}$ × $50m. Thus statement (1) alone is sufficient to answer the question.

Statement (2) above also enables us (along with the pie chart) to compute the total steel consumption and therefore that of the automotive sector. Thus these two statements individually allow us to answer the question. The correct answer is (D).

4. What is the second term of the sequence?
   (1) The third term is 9.
   (2) The second term is 2 more than the first and the third term is 2 more than the second.

Solution: From statement (1) alone all we know is that the third term is 9. We can have several sequences where the third term is 9, such as

1	5	9	13	17
3	6	9	12	15
2	5	9	14	20

Thus statement (1) alone does not give us a unique second term. Now let us examine statement (2) alone. Statement (2) says that there is a difference of 2 between subsequent terms, but this information can also lead to various sequences such as

1	3	5	7	9
2	4	6	8	10
5	7	9	11	13

which does not give a unique second term. But if we combine the two statements, the first fact we know is that the third term is 9 and we can write the following formula:

		9
1st	2nd	3rd

Bringing in statement (2) which says that the third term is bigger than the second by 2 and the second is bigger than the first by 2 leads to the following

5	7	9
1st	2nd	3rd

which actually specifies the sequence.
The correct answer is (C).

5. Two tanks A and B are cylindrical, with tank A being smaller than tank B.
   How many size A tanks full of water will fill tank B?

   (1) Tank B has a radius of 15 cm and tank A has a radius of 10 cm.
   (2) Tank A has a volume of 190 cm^3 while tank B has a height of 20 cm.

Solution: This problem involves comparison of the volume of two cylindrical tanks. Once the volume of the two tanks can be computed the question can be answered. Let us consider statement (1) alone. It gives us the radii of the bases of the two tanks. In order to calculate the volume of a cylindrical tank we require the height as well as the cross-sectional area, the latter being in turn dependent on the radius of the base. Thus statement (1) alone which gives only the radii of the bases of the two tanks and says nothing about their heights is not sufficient to answer the question.

Statement (2) alone gives the volume of one tank and the height of the other. The height alone does not enable us to compute the volume of the second tank. As a result, statement (2) alone is also not sufficient. But the two statements combined enable us to answer the question. The correct answer is (C).

6. Is $S > T$?

   (1) $T^2 > S^2$
   (2) $S - T > 0$

Solution: Statement (1) alone is insufficient to determine whether $S > T$ because it says nothing about the signs of S and T. Suppose $S^2 = 16$ and $T^2 = 9$ then it is true that $S^2 > T^2$ but as to whether $S > T$ or not we cannot be certain because $\sqrt{S^2} = \sqrt{16} = 4$ or -4, and $\sqrt{T^2} = \sqrt{9} = +3$ or -3. If $S = 4$, and $T = 3$ or -3, then $S > T$ but if $S = -4$ and $T = 3$ or -3, then $S < T$.
Thus we cannot say for certain if $S > T$ or $S < T$ even if $S^2 > T^2$.

Statement (2) implies that S > T, as adding T to both sides of the equation leads to this result. Interestingly this is also what the question seeks to determine, so statement (2) alone answers the question.
The correct answer is (B).

7. What is the area of triangle ABC in Figure 2?

   (1) $25 + a^2 = 169$
   (2) $x = 90$

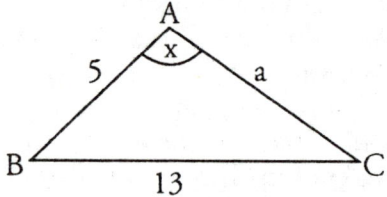

Figure 2

Solution: Statement (1) is one of the common Pythagorean triads, which implies that angle x is 90°. The area of a triangle is half times the base times the height. In this case if AB (5) is the base, then AC (a) is the height. From the Pythagorean relationship, 'a' must be 12 ($13^2 = 5^2 + 12^2$). Thus the area of the triangle equals $\frac{1}{2}$ × 5 × 12 = 30, and statement (1) alone enables us to answer the question.

Statement (2) says that angle x is 90°.
If $x = 90$, then triangle ABC is a right-angled triangle and side AC (a) can be computed.
With side AC computed, the area of the triangle can be subsequently computed. Thus statement (2) alone also enables us to answer the question.
The correct answer is (D).

8. Is *x* a positive odd integer?

    (1) When *x* is divided by 2 there is a remainder of 1.
    (2) *x* is a multiple of 3.

Solution: Statement (1) is sufficient as every odd number divided by 2 leaves a remainder of 1 whereas every even number is exactly divisible by 2.
Statement (2) is not sufficient because there are some multiples of 3 which are even and therefore would not leave a remainder of 1 when divided by 2. Just consider multiples of 3:
3 (3 X 1), 6 (3 X 2), 9 (3 X 3), 12 (3 X 4)
6 and 12 are divisible by 2 whereas 3 and 9 are not. Thus statement (2) alone does not give us a precise answer.
The correct answer is (A).

9. What was the ratio of green to red apples sold by Apple Dealers Ltd in May 1990?

    (1) The company bought equal numbers of both types of apples in April 1990.
    (2) The company sold 1,000 more green apples in May 1990 than in April 1990.

Solution: We are required to determine the ratio of green to red apples sold in May 1990. Statement (1) tells us about apples bought in the previous month, April, but says nothing about May sales which is what we really want. So statement (1) alone does not offer any help to the solution.

Statement (2) tells us that 1,000 more green apples were sold in May than in April. But we do not know what the April sales were. If we did we could add the number 1,000 to determine the number of green apples sold in May. As it is, there is insufficient information in both statements to enable us to compute a ratio.
The correct answer is (E).

10. Are the two triangles in Figure 3 similar?

(1) ∠ n = ∠ m
(2) ∠ x = ∠ y

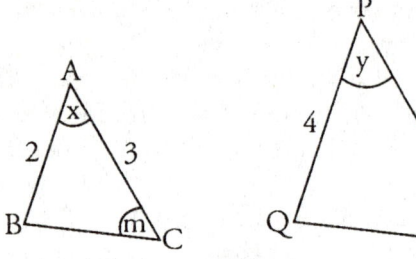

Figure 3

Solution: Two triangles are similar if they are equiangular or their sides are proportional or two sides are proportional and their included angles are equal.
From the figure, the ratio of AB to PQ is equal to the ratio of AC to PR. Statement (1) says that ∠ m = ∠ n but since neither ∠ m nor ∠n are included angles, we cannot say for sure from statement (1) alone that the triangles are similar.
On the other hand statement (2) says ∠ x = ∠ y and since these are included angles, this statement with the information given in the figure satisfies the requirement for similarity. The correct answer is (B).

Caution: Combining statements (1) and (2) will mean that the two triangles are equiangular and therefore similar (if two angles in two triangles are equal then the third angles are also equal). Thus there is the strong temptation to go for option (C). But remember that this section tests your ability to determine when sufficient information to solve a problem is reached. You are first required to consider the two statements separately and then to go on and consider them jointly when each on its own is insufficient to solve the problem. In this case statement (2) alone enables us to solve the problem and the test is to see if you recognise that.

11. What is the value of a two-digit number y?

   (1) The product of the two digits is 10.
   (2) The sum of the two digits is 7.

Solution: The factors of 10 are: 10 and 1, 5 and 2. Thus the number can only be composed of 5 and 2. The possibilities are 52 and 25. Since we have two possibilities, statement (1) alone does not lead to a definite answer. Statement (2) also leads to 5 and 2, 1 and 6, 3 and 4 and to the same situation as above. Thus we require further information. The correct answer is (E).

Caution: Note that combining (1) and (2) may tempt you to choose (C) since 5 and 2 are common. But 5 and 2 may lead to 52 or 25. We don't know the unique answer.

12. What is the volume of a rectangular box?

   (1) The total surface area is 300 cm².
   (2) The width of the box is 5 cm.

Solution: The volume of a rectangular box is given by length (L) × width (W) × height (H). The surface area is given by 2HW + 2LW + 2LH.

Statement (1) says that this is 300 cm² but says nothing about L, W and H. Thus this statement alone does not enable computation of the volume. Statement (2) only gives the width and is also not sufficient. Combining the two statements still leaves us with unresolved values of L and H.

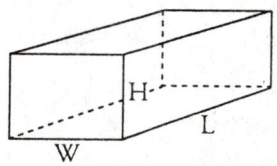

We can go no further in the computation of the volume. Thus we require further information to solve the problem. The correct answer is (E).

13. Which of the following four numbers a, b, c and d is the largest?

    (1) The average of a, b, c and d is 20.
    (2) The numbers a, b and c are each less than 19.

Solution: Statement (1) is insufficient since any of the four numbers could be the largest.
Statement (2) alone is also insufficient since d could be larger than 19 or could be smaller than one of the numbers a, b, or c.
Statement (1) and statement (2) together are sufficient. Statement (1) implies that a + b + c + d = 80 or d = 80 −a −b −c. Now using statement (2) we can see that the smallest value of d is 80 − 19 −19 −19 = 23. So d must be the largest number. The correct answer is (C).

14. If M is an integer, is $M/2$ an even integer?

    (1) M is a multiple of 2^2.
    (2) 2 is a factor of M.

Solution: Statement (1) implies that $4(2^2)$ is a factor of M. As a result M can be written as K x 4. K x $\frac{4}{2}$ equals K x 2, which is even. Thus statement (1) alone allows solution of the problem.
Statement (2) implies that M can be written as P x 2 where P is the result obtained by dividing M by 2. But P can be even or odd, such as 4 or 3 in which case M is 8 or 6. Statement (2) does not lead to a definite answer.

The correct answer is (A).

15. Is $xy > 0$?

    (1) $x^3y^2 > 0$
    (2) $x^2y < 0$

Solution: Note that the question can be reframed thus: Is xy positive?

If $xy > 0$, then both x and y must be positive or both must be negative (recall that the product of two positive numbers or two negative numbers is positive).

Referring to statement (1), x must be positive since y^2 is always positive regardless of the sign of y. But we cannot be sure of y. Thus we cannot answer the problem based on statement (1) above.

Let us now consider statement (2). This indicates that y is definitely negative since x^2 must be positive. Even so we do not know whether x itself is positive or negative. Neither statement (1) nor statement (2) individually is sufficient to answer the question.

Now let us combine the two statements. Statement (1) definitely implies that x is positive and statement (2) implies that y is negative, thus the two statements together enable us to answer the question.

The correct answer is (C).

16. Triangle ABC is isosceles, with AB = BC. What is the area of triangle ABC?

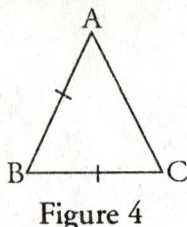

Figure 4

(1) Angle BAC = 60°.
(2) Side AB = 6 inches.

Solution: Statement (1) implies that triangle ABC is equiangular but the lengths of its sides are not known. Since we require lengths to calculate area, statement (1) alone is not sufficient.
Statement (2) alone implies that side BC is also 6 inches but side AC is not known, neither is any angle that may lead to its determination. Thus statement (2) alone is also not sufficient. But the two statements combined indicate that $\triangle ABC$ is equiangular with sides 6 in. This enables us to calculate the height of the triangle, and thus its area. The correct answer is (C).

17. What is the value of $x - y$?

(1) $y - x = x - y$
(2) $y^2 - x^2 = y - x$

Solution: From statement (1) we can obtain $2y = 2x$ or $y = x$. Thus $(x - y)$ is zero. Thus statement (1) alone leads to a solution of the problem.
Let us check statement (2).
$y^2 - x^2 = y - x$, can be written as $(y + x)(y - x) = y - x$ or $(y - x)(y + x - 1) = 0$
That is, $y = x$ or $y + x + 1 = 0$ and this is not definite. The correct answer is (A).

18. How many more miles does car A cover than car B if they both travel for 3 hours?

    (1) The speed of car A is 50 mph.
    (2) The speed of car B is 10 mph less than that of car A.

Solution: Statement (1) implies that car A covers a distance of 50 x 3 = 150 miles but says nothing of the speed and thus the distance covered by car B. Thus statement (1) alone is not sufficient.

Statement (2) alone does not give the actual speed of car B or car A.
It is thus not sufficient. A combination of the two statements, however, leads to a solution of the problem. From statement (1) we can calculate the distance covered by car A. From statement (2) we know that the speed of car B is 40 mph and consequently the distance it covered is 40 x 3 = 120 mph. The difference between the two distances is the required information.
The correct answer is (C).

19. If $y = x^2$, what is the value of $y - x$?

    (1) $x = 2$
    (2) $x + y = 6$

Solution: From statement (1) $y = 4$ and the value of $(y - x)$ can easily be determined. We can rewrite statement (2) as $x + x^2 = 6$, which has 2 values of x. Thus there is no unique $(y - x)$.
The correct answer is (A).

20. What percentage of an MBA class is women?

   (1) The class has 30 more men than women.
   (2) The total number of students is 150.

Solution: If there are x women in the class then statement (1) implies that there are $x + 30$ men and a total class of $x + (x + 30)$. The percentage of women in the class is given by

$$\frac{x}{x + (x + 30)} \times 100$$

This cannot be determined and therefore statement (1) alone is not sufficient. Statement (2) alone does not give the number of either women or men but just the total class size and is also not sufficient. But here too the combination of both statements leads to a solution. Let, as above, the number of women be x. Then from statement (1) the number of men is $(x + 30)$. Since from statement (2) the total class size is 150 it follows that $x + (x + 30) = 150$, from which $x = 60$. Thus the ratio of women to the total class is $\frac{60}{150}$ or 40 per cent.

The correct answer is (C).

21. If x, y and z are the lengths of the three sides of a triangle, is $x > z$?

   (1) $x = 5$
   (2) $y = 6$ and $z = 7$

Solution: The key to solving the problem is recognition of the fact that the sum of the lengths of any two sides of a triangle is always greater than the third side. Thus $x + y > z$.
Statement (1) implies that $y + 5 > z$ and does not lead to a solution.
Statement (2) implies that $6 + x > 7$, from which $x > 1$. This has countless values of x. Both statements together give $x = 5$, $y = 6$ and $z = 7$. We can thus answer the question. The correct answer is (C).

22. If both x and y are non-zero numbers, what is the value of
$$\frac{x^2 y^4 (xy)^2}{y^3 x^4}$$

   (1) x = 2
   (2) y = 4

Solution: Rather than rushing into this without a thought and choosing option (C), it is important to realise that simplifying this expression should be the first step.

$$\frac{x^2 y^4 (xy)^2}{y^3 \times x^4} = \frac{x^2 y^4 x^2 y^3}{y^3 \times x^4} = y^3$$

Thus only statement (2) is required to evaluate the expression (which is actually just y^3). Statement (1) is not required at all.
The correct answer is (B).

23. What is the difference in volume between sphere A and sphere B?

   (1) The ratio of the volume of sphere A to the volume of sphere B is 2.
   (2) The radius of sphere B is 4 cm.

Solution: The volume of a sphere is given by $\frac{4\pi r^3}{3}$, where r is the radius.
From statement (1) we know that if the radius of sphere A is r_a and that of sphere B is r_b,

$$\frac{\frac{4\pi r_a^3}{3}}{\frac{4\pi r_b^3}{3}} = 2 \quad \text{or} \quad \frac{r_a^3}{r_b^3} = 2$$

Since we require the difference of volumes we need to know the radius of at least one of the spheres. Statement (1) does not provide this and is not therefore sufficient.

Statement (2) on its own gives only the radius of sphere B and by implication the volume of sphere B but says nothing about the volume of sphere A. Thus statement (2) alone is also not sufficient. The two statements together however lead to a solution. With one radius given we can derive the radius of sphere A from the relationship

$$\frac{r_a^3}{r_b^3} = 2$$

Thus the volume of both spheres can be calculated and hence the difference in volume.
The correct answer is (C).

24. How many seconds are there in time period T?

   (1) Period T is 50 minutes long.
   (2) Period T starts at 11.37pm and ends at 12.27am.

Solution: This is a very easy question. As has already been mentioned the GMAT is composed of easy, slightly difficult and difficult questions. There should be no time wasting on such easy questions. Any time saved then goes into more difficult questions. All we need is the length of the time period in minutes, hours, days, weeks etc. and we can compute the number of seconds in it. As can be seen, statement (1) gives the length of T as 50 minutes which is 50 × 60 seconds.

Do note that very easy questions also can be tricky. Statement (2) tempts us to think that the time period is from 11.37pm one day to 12.27am the following day. That is plausible but the problem does not say the times given are for consecutive days. Thus statement (2) is not definite.
The correct answer is (A).

25. Is DC parallel to AB in the figure shown?

    (1) AB = AC
    (2) BC = AC

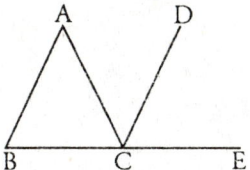

Figure 5

Solution: Statement (1) indicates that triangle ABC is an isosceles triangle with angle ABC equal to angle ACB. It does not say anything about line DC or the angle ACD. It is therefore not sufficient.
Statement (2) only also implies that triangle ABC is an isosceles triangle with angle CBA equal to angle CAB. It is also not sufficient. The two statements together imply that triangle ABC is equiangular but say nothing about either angle ACD or angle DCE. Thus more information is required to answer the question.
The correct answer is (E).

26. Is x bigger than y?

    (1) x is bigger than z
    (2) z is not as big as y

Solution: Statement (1) does not say anything about y and is therefore not adequate.

Statement (2) does not relate to x, and is therefore not adequate either. Both statements together tell us that both x and y are bigger than z but give no indication as to how x and y relate to each other. We therefore cannot answer the question without further information.
The correct answer is (E).

27. What is the area of the shaded portion of the quadrilateral ABCD?

   (1) The area of ABCD is 16 cm².
   (2) ABCD is a square.

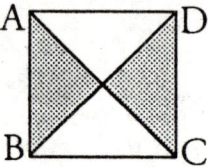

Figure 6

Solution: Statement (1) is not sufficient. Unless ABCD is a rectangle or parallelogram, the area of the shaded portion will depend on the relative lengths of AB and AD. Since we do not know what the actual figure is we cannot determine the answer to the question.

Statement (2) is also not sufficient. We are informed that ABCD is a square but without dimensions we cannot determine area.

Statements (1) and (2) combined, however, offer a solution. Since ABCD is a square (from statement (2)), the shaded portion is half the area of the square. And since the area of the square is 16 cm² (from statement (1)), the shaded portion must have an area of 8 cm². The correct answer is (C).

28. If x and y are non-zero positive integers, is $\frac{y}{x}$ an integer?

    (1)   There is only one pair of positive integers whose product equals x.
    (2)   y is the product of 2 and some other integer.

Solution: Both statements are obviously not sufficient, even when combined.

The quotient $\frac{y}{x}$ is not unique and can be an integer or a fraction.

The correct answer is (E).

29. If N is an integer between 2 and 90 and N is also the square of an integer, what is the value of N?

    (1)   N is the cube of an integer.
    (2)   N is even.

Solution: Let us put down the facts.
N is an integer between 2 and 90.
N is also the square of an integer.
Perfect squares between 2 and 90 are:
4, 9, 16, 25, 36, 49, 64, 81.
Statement (1) points only to 64, which is between 2 and 90, is a square of an integer (8) and is a cube of an integer (4).
Statement (2), does not help. There are many even number squares between 2 and 90, ie 4, 16, 36, 64.
The correct answer is (A).

30. What was the original price of a lap-top computer which was sold for $1900?

    (1)   The sales tax on the computer was 20%.
    (2)   The computer was sold at a discount of 5%.

Solution: Statement (2) implies that the original price was $2000.
If $0.95 \times$ price $= \$1900$, then the price is $2000.
Statement (2) is sufficient.
Statement (1) is not sufficient. The sales tax is always 20% regardless of the price charged to the buyer.
The correct answer is (B).

PART 3
Four Part Practice Examination

ANSWER SHEET

Shade the ovals which correspond to your answers.

Practice Exam Section 1: Problem Solving

1. Ⓐ Ⓑ Ⓒ Ⓓ Ⓔ   5. Ⓐ Ⓑ Ⓒ Ⓓ Ⓔ   9. Ⓐ Ⓑ Ⓒ Ⓓ Ⓔ   13. Ⓐ Ⓑ Ⓒ Ⓓ Ⓔ
2. Ⓐ Ⓑ Ⓒ Ⓓ Ⓔ   6. Ⓐ Ⓑ Ⓒ Ⓓ Ⓔ   10. Ⓐ Ⓑ Ⓒ Ⓓ Ⓔ   14. Ⓐ Ⓑ Ⓒ Ⓓ Ⓔ
3. Ⓐ Ⓑ Ⓒ Ⓓ Ⓔ   7. Ⓐ Ⓑ Ⓒ Ⓓ Ⓔ   11. Ⓐ Ⓑ Ⓒ Ⓓ Ⓔ   15. Ⓐ Ⓑ Ⓒ Ⓓ Ⓔ
4. Ⓐ Ⓑ Ⓒ Ⓓ Ⓔ   8. Ⓐ Ⓑ Ⓒ Ⓓ Ⓔ   12. Ⓐ Ⓑ Ⓒ Ⓓ Ⓔ   16. Ⓐ Ⓑ Ⓒ Ⓓ Ⓔ

Practice Exam Section 2: Problem Solving

1. Ⓐ Ⓑ Ⓒ Ⓓ Ⓔ   5. Ⓐ Ⓑ Ⓒ Ⓓ Ⓔ   9. Ⓐ Ⓑ Ⓒ Ⓓ Ⓔ   13. Ⓐ Ⓑ Ⓒ Ⓓ Ⓔ
2. Ⓐ Ⓑ Ⓒ Ⓓ Ⓔ   6. Ⓐ Ⓑ Ⓒ Ⓓ Ⓔ   10. Ⓐ Ⓑ Ⓒ Ⓓ Ⓔ   14. Ⓐ Ⓑ Ⓒ Ⓓ Ⓔ
3. Ⓐ Ⓑ Ⓒ Ⓓ Ⓔ   7. Ⓐ Ⓑ Ⓒ Ⓓ Ⓔ   11. Ⓐ Ⓑ Ⓒ Ⓓ Ⓔ   15. Ⓐ Ⓑ Ⓒ Ⓓ Ⓔ
4. Ⓐ Ⓑ Ⓒ Ⓓ Ⓔ   8. Ⓐ Ⓑ Ⓒ Ⓓ Ⓔ   12. Ⓐ Ⓑ Ⓒ Ⓓ Ⓔ   16. Ⓐ Ⓑ Ⓒ Ⓓ Ⓔ

Practice Exam Section 3: Data Sufficiency

1. Ⓐ Ⓑ Ⓒ Ⓓ Ⓔ   6. Ⓐ Ⓑ Ⓒ Ⓓ Ⓔ   11. Ⓐ Ⓑ Ⓒ Ⓓ Ⓔ   16. Ⓐ Ⓑ Ⓒ Ⓓ Ⓔ
2. Ⓐ Ⓑ Ⓒ Ⓓ Ⓔ   7. Ⓐ Ⓑ Ⓒ Ⓓ Ⓔ   12. Ⓐ Ⓑ Ⓒ Ⓓ Ⓔ   17. Ⓐ Ⓑ Ⓒ Ⓓ Ⓔ
3. Ⓐ Ⓑ Ⓒ Ⓓ Ⓔ   8. Ⓐ Ⓑ Ⓒ Ⓓ Ⓔ   13. Ⓐ Ⓑ Ⓒ Ⓓ Ⓔ   18. Ⓐ Ⓑ Ⓒ Ⓓ Ⓔ
4. Ⓐ Ⓑ Ⓒ Ⓓ Ⓔ   9. Ⓐ Ⓑ Ⓒ Ⓓ Ⓔ   14. Ⓐ Ⓑ Ⓒ Ⓓ Ⓔ   19. Ⓐ Ⓑ Ⓒ Ⓓ Ⓔ
5. Ⓐ Ⓑ Ⓒ Ⓓ Ⓔ   10. Ⓐ Ⓑ Ⓒ Ⓓ Ⓔ   15. Ⓐ Ⓑ Ⓒ Ⓓ Ⓔ   20. Ⓐ Ⓑ Ⓒ Ⓓ Ⓔ

Practice Exam Section 4: Data Sufficiency

1. Ⓐ Ⓑ Ⓒ Ⓓ Ⓔ   6. Ⓐ Ⓑ Ⓒ Ⓓ Ⓔ   11. Ⓐ Ⓑ Ⓒ Ⓓ Ⓔ   16. Ⓐ Ⓑ Ⓒ Ⓓ Ⓔ
2. Ⓐ Ⓑ Ⓒ Ⓓ Ⓔ   7. Ⓐ Ⓑ Ⓒ Ⓓ Ⓔ   12. Ⓐ Ⓑ Ⓒ Ⓓ Ⓔ   17. Ⓐ Ⓑ Ⓒ Ⓓ Ⓔ
3. Ⓐ Ⓑ Ⓒ Ⓓ Ⓔ   8. Ⓐ Ⓑ Ⓒ Ⓓ Ⓔ   13. Ⓐ Ⓑ Ⓒ Ⓓ Ⓔ   18. Ⓐ Ⓑ Ⓒ Ⓓ Ⓔ
4. Ⓐ Ⓑ Ⓒ Ⓓ Ⓔ   9. Ⓐ Ⓑ Ⓒ Ⓓ Ⓔ   14. Ⓐ Ⓑ Ⓒ Ⓓ Ⓔ   19. Ⓐ Ⓑ Ⓒ Ⓓ Ⓔ
5. Ⓐ Ⓑ Ⓒ Ⓓ Ⓔ   10. Ⓐ Ⓑ Ⓒ Ⓓ Ⓔ   15. Ⓐ Ⓑ Ⓒ Ⓓ Ⓔ   20. Ⓐ Ⓑ Ⓒ Ⓓ Ⓔ

Section 1: Problem Solving

16 questions, time allowed 25 minutes

Directions:
In this section solve each problem, using any available space on the page for rough work. Then indicate the best of the answer choices given.

Numbers:
All numbers used are real numbers.

Figures:
Figures that accompany problems in this section are intended to provide information useful in solving the problems. They are drawn as accurately as possible except when it is stated in a specific problem that its figure is not drawn to scale. All figures lie in a plane unless otherwise indicated.

1. 10 per cent of the students who entered Pasco University for a 3 year programme in 1987 did not graduate in 1990. If 1100 students entered Pasco as freshmen in 1987 and 1100 students in total graduated in 1990, how many of these students did not enter in 1987?

   (A) 100   (B) 110   (C) 120   (D) 130   (E) 140

2. Which of the following has a value greater than 1?

   (A) $\frac{\sqrt{2}}{2}$ (B) $\frac{2}{\sqrt{2}}$ (C) $\left(\frac{5}{6}\right)^2$ (D) $4\left(\frac{2}{9}\right)$ (E) $2\left(\frac{3}{4}\right)^3$

3. There are three types of Wimbledon tickets: Centre Court, which cost $25, West End which cost $18 and East End which cost $15. If x Centre Court, y West End and z East End tickets are sold, what percentage of total ticket revenue is West End?

   (A) $\dfrac{1800y}{25x+18y+15z}$ (B) $\dfrac{18y}{25x+18y+15z}$ (C) $\dfrac{15z}{25x+18y+15z}$

   (D) $\dfrac{25y}{25x+18y+15z}$ (E) $\dfrac{0.18y}{25x+18y+15z}$

4. A part-time sales girl earns $500 per month plus 5 per cent of whatever total sales she makes. If in 1990 she achieved $20,000 sales how much pay did she earn that year?

   (A) $5000   (B) $6,000   (C) $7,000   (D) $7500   (E) $8,000

5. If $\dfrac{x^2+x-3}{3} = 1$, x could equal

   (A) 3   (B) 2   (C) –2   (D) 1   (E) 6

6. Two fast trains A and B are travelling on parallel tracks. A is travelling at 160 km/hr while B is travelling at 200 km/hr. At 10am train A is 200 km ahead of train B. At what time does train B catch up with train A?

    (A) 1pm (B) 2pm (C) 3pm (D) 4pm (E) 5pm

7. ABCD is a square of side 10 inches with an inscribed circle as shown in Figure 1. Which is the closest approximation to the area of the shaded portion?

    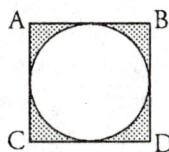

    Figure 1

    (A) 15 in² (B) 31 in² (C) 25 in² (D) 41 in² (E) 21 in²

8. The number 40 is 12.5 per cent of which of the following:

    (A) 400 (B) 480 (C) 520 (D) 320 (E) 300

9. Which of the following is equal to 351/558?

    (A) $\frac{109}{171}$ (B) $\frac{117}{206}$ (C) $\frac{39}{62}$ (D) $\frac{7}{39}$ (E) $\frac{39}{89}$

10. The price of a train ticket in a certain country bears a direct relationship with the length of the journey. If the ticket for a 500 km journey is $80, how much is charged for a journey of y km?

    (A) $\frac{50y}{8}$ (B) $\frac{8y}{50}$ (C) $\frac{8}{50y}$ (D) $\frac{400}{y}$ (E) 400y

11. If $\frac{x}{21}$ is 1 more than $\frac{y}{21}$, then y equals

    (A) x + 1 (B) x − 1 (C) x + 21 (D) x − 21 (E) x

115

12. If n and p are odd integers and m is an even integer, which of the following *cannot* be an integer?

   (A) $\frac{m}{p}$ (B) $\frac{n}{m}$ (C) $\frac{n}{p}$ (D) $\frac{mn}{p}$ (E) $\frac{mp}{n}$

13. A man had his first son when he was 24. If the son is now one-third of the man's age, how old is the man?

   (A) 30   (B) 36   (C) 40   (D) 42   (E) 48

14. A box of 21 packets of crisps costing 320 cents contains 3 different flavours, prawn, vinegar and beef. These cost 10, 15 and 20 cents respectively. The box contains twice as many packets of prawn flavour crisps as vinegar flavour. How many packets of beef flavour crisps are in the box?

   (A) 12   (B) 10   (C) 11   (D) 7   (E) 9

15. When a positive integer n is divided by 15 the remainder is 5 and the quotient is K. When n is divided by 19 the quotient is m and the remainder is 12. Which of the following is true?

   (A) 15K + 19m = 17   (B) 15K = 19m + 17
   (C) K + m = 5 + 12   (D) 15K − 19m = 7
   (E) 19K + 15m = 60

16. In Figure 2, 3 squares and a triangle are arranged as shown with areas P, Q, R and S. If the area of P = 16, the area of Q = 9 and the area of R = 25, what is the area of S?

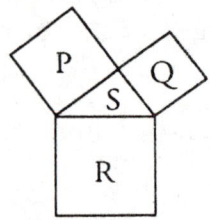

Figure 2

   (A) 5   (B) 6   (C) 7   (D) 8   (E) 9

Section 2: Problem Solving

16 questions, time allowed 25 minutes

Directions:
In this section solve each problem, using any available space on the page for rough work. Then indicate the best of the answer choices given.

Numbers:
All numbers used are real numbers.

Figures:
Figures that accompany problems in this section are intended to provide information useful in solving the problems. They are drawn as accurately as possible except when it is stated in a specific problem that its figure is not drawn to scale. All figures lie in a plane unless otherwise indicated.

1. If $\frac{3}{2}(27 \div \frac{9}{8}) = x$, what is x?

   (A) $\frac{27}{16}$ (B) $\frac{729}{16}$ (C) 39   (D) 36   (E) 32

2. If y is an integer which of the following must be an odd number?

   (A) $3y+2$ (B) $9y$ (C) $4y+5$ (D) y^2 (E) y^3

3. An electron microscope allows a scientist to see an organism 2,000 times its actual size. If the diameter of a certain organism appears to be 0.5 cm, what is the actual diameter of the organism?

   (A) 25×10^{-5} cm   (B) 5×10^{-5} cm   (C) 25×10^{-3} cm
   (D) 5×10^{-3} cm   (E) 5×10^{-4} cm

4. If one of the roots of the equation $2x^2 + 5x - q = 0$ is -3, what is the value of q?

   (A) 1   (B) 2   (C) 3   (D) 4   (E) 5

5. For a certain scientific experiment a spherical balloon was blown up until its diameter reached 10 ft. If it was determined that this volume was inadequate and the balloon was blown up further until its diameter reached 12 ft, by what proportion of the first volume did the volume increase?

   (A) 20%   (B) 50%   (C) 60.8%   (D) 72.8%   (E) 80.6%

6. If A, B and C share a gift of $24,300 in the ratio of 15:10:2 respectively, how much does B get?

   (A) 4,500   (B) 7,500   (C) 9,000   (D) 10,500   (E) 1,200

7. Rail freight costs $y per tonne per km. How much does it cost to transport 9 cars each of which weighs 1.2 tonnes for x km?

   (A) 10.8xy (B) $\frac{10.8x}{y}$ (C) $\frac{x}{10.8y}$ (D) 1.2xy (E) 9xy

8. A 1,000 tablet pack of Zantac costs $12 while a 500 tablet pack of the same drug costs $7. By how much is a tablet in the larger pack cheaper than one in the smaller pack?

   (A) 1 cent (B) 0.8 cent (C) 0.5 cent (D) 0.2 cent (E) 0.1 cent

9. If n is an integer greater than 1, which of the following has the greatest value?

   (A) $\frac{1}{(n-1)^2}$ (B) $\frac{1}{n^2-1}$ (C) $\left(\frac{1}{n-1}\right)^2$ (D) $\frac{1}{n-n^2}$ (E) $\frac{1}{n-1}$

10. In a certain hospital 70 per cent of the nurses are female and 60 per cent of these female nurses wear blue uniforms. If 50 per cent of all the nurses wear blue uniforms, what per cent of the nurses who do not wear blue uniforms are male?

    (A) 28 (B) 32 (C) 44 (D) 20 (E) 14

11. 81 cubic inches of water is poured into a tank with a square base. If the height the water reaches is three times the width of the tank what is the area of the base?

    (A) 3 (B) $2\sqrt{6}$ (C) 9 (D) 12 (E) 10

12. If $x>0$, which of the following is equal to $\sqrt{32x}$?

    (A) $2\sqrt{2x}$ (B) $4\sqrt{2x}$ (C) $3\sqrt{2x}$ (D) $6\sqrt{x}$ (E) $4\sqrt{x}$

13. In figure 1, AC = 4 ft, ∠ABC = 60° and ∠BCD = 120°. What is the area of triangle ABC?

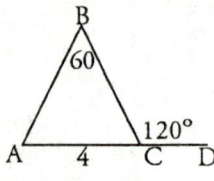

Figure 1

(A) √6   (B) 2√6   (C) 2√3   (D) 4√3   (E) 6

14. How many litres of pure salt must be added to a 100 litre solution that is 20 per cent brine in order to produce a solution that is 30 per cent brine?

(A) 30.0   (B) 24.2   (C) 20.0   (D) 15.0   (E) 14.3

15. A manufacturer believes that if he makes and sells n units of a product his sales revenue will be 20 per cent greater than his total costs. If he actually only sold 90 per cent of the n units he produced, by what percentage was the total revenue greater than the total costs of production?

(A) 25   (B) 20   (C) 15   (D) 10   (E) 8

16. If S is an integer, which of the following could be an odd integer?

(A) 2S   (B) 6S   (C) 8S   (D) 9S   (E) 18S

Section 3: Data Sufficiency

20 questions, time allowed 25 minutes

Directions
Each of the data sufficiency problems below consists of a question and two statements labelled (1) and (2), which contain some information. You have to decide whether the data or information given in the two statements is sufficient for answering the question.

Using the data given in the statements plus your knowledge of mathematics and everyday facts (such as the number of days in April or the meaning of terms such as clockwise), you choose option
(A) if statement (1) ALONE is sufficient, but statement (2) ALONE is not sufficient to answer the question asked;
(B) if statement (2) ALONE is sufficient, but statement (1) ALONE is not sufficient to answer the question asked;
(C) if both statements (1) and (2) TOGETHER are sufficient to answer the question asked, but NEITHER statement ALONE is sufficient;
(D) if EACH statement alone is sufficient to answer the question asked; and
(E) if statements (1) and (2) together are not sufficient to answer the question asked and additional data specific to the problem is needed.

1. If a Ford car sells for $8,000 today, how much will it sell for in 2 years time?

   (1) The price of the car will rise by 10 per cent per annum during this 2 year period.
   (2) The price of the car in 2 years time will be 1.21 times its price today.

2. What is the value of $y - z$?

   (1) $y + z = 4$
   (2) $y - z < 4$

3. By what fraction has the price of a gold necklace reduced?

   (1) The original price was $260.
   (2) The new price is $13 less than the original.

4. A rectangle has length L and width W. What is the length L?

   (1) The ratio of the length to the width is 3 to 2.
   (2) The perimeter of the rectangle is 192 metres.

5. What is the average (arithmetic mean) of 6 numbers?

   (1) 3 of the numbers are greater than 10.
   (2) The smallest number is 5 and the biggest is 20.

6. Is a^2 equal to ab?

   (1) $a = b$
   (2) $(a+b)^2 = a^2 + b^2 + 2ab$

7. In the parallelogram PQRS, what is the measure of angle PQR?

   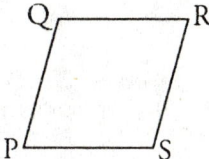

   Figure 1

   (1) Angle PSR is obtuse.
   (2) Angle SPQ is 70°.

8. In triangle ABC what is the length of BC?

   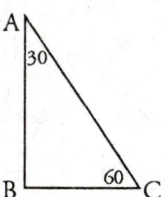

   Figure 2

   (1) $AC = 12$
   (2) $AB = 6\sqrt{3}$

9. Is n an integer greater than 5?

   (1) 4n is a positive integer.
   (2) $\frac{n}{6}$ is a positive integer.

10. Which of 2 plants produces a product at lower unit cost? Plant A has a cost function given by 1,000 + 2Q while plant B has a cost function given by 100 + Q^2, where Q is the level of output?

    (1) Q is 1,000.
    (2) Plant A was designed and built in 1960 while plant B was built in 1980.

11. 50 per cent of those who take part in a test marketing exercise promise to buy the product. What percentage of those who take part in the exercise actually buy the product?

    (1) 75 per cent of those who buy receive a gift.
    (2) 20 per cent of those who promise to buy actually buy the product but nobody else does?

12. Is r–s > x–y?

    (1) x > r and y < s
    (2) y = 4, s = 8, r = 6 and x = 7

13. What is the difference in the distance covered by Nigel and Freydis?

    (1) Nigel drives twice as fast as Freydis.
    (2) Freydis drove for 6 hours whilst Nigel drove for 4 hours.

14. A car travels for 20 minutes on a level track. What is its average speed in km/hr?

    (1) The car travelled a total of 15 km.
    (2) The car's minimum speed was 35 km/hr.

15. If y is a positive integer, what is the value of y?

    (1) $y^2 - 6y + 9 = 0$
    (2) $\frac{1}{4} < \frac{1}{y} < \frac{1}{2}$

16. A retailer sells an item and makes 40 per cent profit. How much did the retailer pay for the item in the first place?

    (1) The retailer sold the item for $1,200.
    (2) If the retailer had sold the item at a 10 per cent discount he would have made only 26 per cent.

17. If n is a positive number less than 10 and N = 421 -- n, what is the value of n?

    (1) N is divisible by 7.
    (2) N is divisible by 3.

18. Is quadrilateral ABCD a parallelogram?

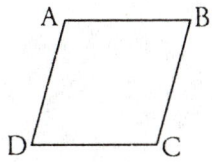

Figure 3

    (1) $\angle A = \angle C$
    (2) DC is parallel to AB.

19. Is a^2-b^2 a positive number?

   (1) (a–b) is a positive number.
   (2) (a+b) is a positive number.

20. Is the integer N odd?

   (1) N divided by 5 leaves a remainder of 1.
   (2) N divided by 8 leaves a remainder of 1.

Section 4: Data Sufficiency

20 Questions, time allowed 25 minutes

Directions
Each of the data sufficiency problems below consists of a question and two statements labelled (1) and (2), which contain some information. You have to decide whether the data or information given in the two statements is sufficient for answering the question.

Using the data given in the statements plus your knowledge of mathematics and everyday facts (such as the number of days in April or the meaning of terms such as clockwise), you choose option
(A) if statement (1) ALONE is sufficient, but statement (2) ALONE is not sufficient to answer the question asked;
(B) if statement (2) ALONE is sufficient, but statement (1) ALONE is not sufficient to answer the question asked;
(C) if both statements (1) and (2) TOGETHER are sufficient to answer the question asked, but NEITHER statement ALONE is sufficient;
(D) if EACH statement alone is sufficient to answer the question asked; and
(E) if statements (1) and (2) together are not sufficient to answer the question asked and additional data specific to the problem is needed.

1. By what proportion is the area of a rectangle increased?

   (1) The length is increased by 20 per cent.
   (2) The width is decreased by 10 per cent.

2. If $x + y = 6$, what is y?

   (1) $\frac{y}{x} = 2$

   (2) $x^2 - 4x + 4 = 0$

3. How much does it cost to make 50 copies of a map?

   (1) It costs $1.00 for the first 10 copies.
   (2) It costs 10 cents per copy.

4. What is the distance from New York to Boston?

   (1) A car travelling at 50 mph covers the distance in 5 hours.
   (2) A plane flying at 500 mph covers the distance in half an hour.

5. Is $xy < 0$?

   (1) $\frac{x}{y} < 0$

   (2) $x - y < 0$

6. How many more cars does factory x produce than factory y?

   (1) Factory x turns out half as much again as factory y every year.
   (2) Factory y turns out 10,000 cars every year.

7. Is P > 1?

   (1) P + M > 4
   (2) M < 3

8. If $2x + 2xy + z = 2xy + 6$, what is the value of x?

   (1) z = 3
   (2) y = 2

9. In the diagram below what is the value of angle x?

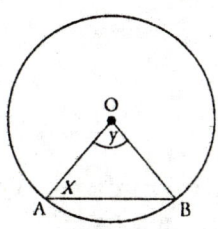

   Figure 1

   (1) y = 60°
   (2) AB = OA

10. How many units of electricity were consumed in December?

    (1) The level of consumption in December was twice that in September.
    (2) The level of consumption in September was half that in June.

11. Is the integer K even?

    (1) $\frac{K}{2}$ is odd.

    (2) 2K is even.

12. What is the area of triangle ABC below?

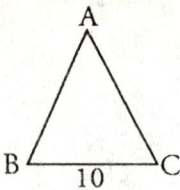

Figure 2

   (1) AB = AC
   (2) AB = 6

13. What is the value of $(x - y)$?

   (1) $2x = 2y + 10$
   (2) $x = 8$

14. A circular pitch was expanded by increasing its diameter by 2 feet. By how much was the surface area increased?

   (1) The original area was 90 square feet.
   (2) The original diameter was 10.7ft.

15. Is the integer k divisible by 15?

   (1) k is divisible by 5.
   (2) k is divisible by 3.

16. In the diagram below, is the line TS parallel to the line QR?

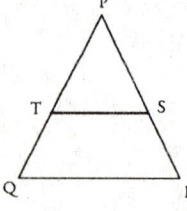

Figure 3

   (1) $\dfrac{PT}{PQ} = \dfrac{PS}{PR}$

   (2) $\angle PTS = \angle PQR$

17. Is y > x?

    (1) $\dfrac{y}{x} = \dfrac{4}{3}$

    (2) $y^2 > x^2$

18. What was Mercy's salary in 1994?

    (1) Mercy's salary increased by 5 per cent each year over the three years to 1994.
    (2) Mercy's salary doubled between 1990 and 1994.

19. How many blue telephones are there in a box?

    (1) 62.5 per cent of the telephones in the box are blue.
    (2) There are 30 telephones in the box that are not blue.

20. How old is John?

    (1) John is one-sixth his father's age.
    (2) John's age is one-sixth the sum of the first 8 positive integers.

ANSWER KEY TO PRACTICE EXAM

Section 1	Section 2	Section 3	Section 4
1 B	1 D	1 D	1 C
2 B	2 C	2 E	2 D
3 A	3 A	3 C	3 B
4 C	4 C	4 C	4 D
5 B	5 D	5 E	5 A
6 C	6 C	6 A	6 C
7 E	7 A	7 B	7 C
8 D	8 D	8 D	8 A
9 C	9 E	9 B	9 D
10 B	10 C	10 A	10 E
11 D	11 C	11 B	11 A
12 B	12 B	12 D	12 C
13 B	13 D	13 E	13 A
14 E	14 E	14 A	14 D
15 D	15 E	15 D	15 C
16 B	16 D	16 A	16 D
		17 C	17 E
		18 C	18 E
		19 C	19 C
		20 B	20 B

Practice Examination Self-Scoring Scale

Section	Poor	Fair	Good	Excellent
1	0-7	8-10	11-14	15-16
2	0-7	8-10	11-14	15-16
3	0-9	10-13	14-17	18-20
4	0-9	10-13	14-17	18-20

A rating of Poor or Fair in any of the Sections of the Practice Examination indicates a lack of basic aptitude or technical competence. If your score falls into this category a thorough review is urgently required.

PART 4

Analysis Of Solutions To Practice Examination

ANALYSIS OF SOLUTIONS

Section 1

Problem Solving

1. The number of students who dropped out of the 1987 class = 10% of 1100 or 110. Thus (1100 − 110) of the 1987 students did actually graduate. But the number of students in the 1990 graduating class was 1100. This figure must be made up of 990 plus 110 students from elsewhere.
The correct answer is (B).

2. A fraction is greater than 1 if the numerator is greater than the denominator.
Note that (C) is a fraction less than 1. Squaring such a fraction makes it even smaller. The same applies to (E). Although the fraction $\frac{3}{4}$ has a coefficient of 2, the fact that it is cubed makes it so small that the factor of two does not offset it. In (D), $4 \times \frac{2}{9}$ is $\frac{8}{9}$ which is less than 1. Thus (E), (C), and (D) can be eliminated leaving (A) and (B). Since $\sqrt{2}$ is less than 2 (equal to 1.414 – you are required to know that $\sqrt{2} = 1.414$ and $\sqrt{3} = 1.732$) (A) is $\frac{1.414}{2}$ which is less than 1 and (B) is $\frac{2}{1.414}$ which is greater than 1 (note that $\frac{2}{2} = 1$). **The correct answer is (B).**

3. Centre Court sales = $x \times \$25 = \$25x$
West End sales = $y \times \$18 = \$18y$
East End sales = $z \times \$15 = \$15z$
Total sales = $\$(25x + 18y + 15z)$
Proportion of West End sales to total sales

$$= \frac{18y}{25x + 18y + 15z} \times 100\%$$

$$= \frac{1800y}{25x + 18y + 15z}$$

The correct answer is (A).

Short cut: Since the answer is a fraction with total sales as a denominator, a quick check of the options is the way to proceed. It is clear that all the denominators are the same, so it does not even matter if you cannot establish what the total sales are, as we did in the conventional solution above. The numerator must be West End sales which is $18y$ and option (B) can be tempting until you recognise that the answer must be a percentage, which leaves (A) as the right option.

4. A salary of $500 per month yields an annual income of $500 × 12 or $6000. To this must be added a bonus equivalent to $\frac{5}{100}$ × $20,000 or $1000. Thus the sales girl earned $7000 in 1990. **The correct answer is (C).**

5. By multiplying both sides of the equation by 3, the equation becomes

$$x^2 + x - 3 = 3$$
or $$x^2 + x - 6 = 0$$

Factorising, $(x + 3)(x - 2) = 0$
giving $x = -3$ and $x = 2$

The correct answer is (B).

Short cut: The method of substituting the values of the various options will be faster if you are not up to speed with factorising quadratic equations. Usually the technique is to substitute the lowest and highest options in the problem to see what comes out. Notice that option (E) can be eliminated because 6^2 will far outweigh -6 in $x^2 + x - 6 = 0$ thus making the left hand side positive rather than zero. Option (A) will do the same. Option (D) will yield a negative number. Option (B) makes the left hand side zero, equal to the right hand side.

6. Let the time required for train B to catch up with train A be x hours.

Then in that period of time
distance travelled by A = $160x$
distance travelled by B = $200x$

But A starts 200 km ahead of B. For B to catch up with A it must cover 200 km plus the $160x$ km B travels in x hours (see below).

```
         |———————————| Distance covered by A
              160x km
|————|————————————————| Distance covered by B
 200 km      160x
|—————————— 200x ——————→ Distance covered by B
```

Thus $200 + 160x = 200x$
giving $x = 5$ hours.

Since the start time is 10am, train B catches up with train A at 3pm (5 hours later). **The correct answer is (C).**

Caution: Be careful not to translate the 5 hours into 5pm thereby choosing option (E).

7. The area of the shaded portion is the difference between the area of the square (10 × 10 sq in) and the area of the circle, πr^2. Since the diameter of the circle is the same as the length one side of the square (10in), the radius of the circle must be 5 in.

Thus area of shaded portion = $10 \times 10 - \pi (5)^2$ (i)
= $100 - 25\pi$ (ii)
= $100 - 78.6$ (iii)
= 21 sq ft in (approx) (iv)

The correct anwser is (E).

Short cut: It is quite straightforward to derive equation (i). What may be time consuming is moving from (ii) to (iii), i.e. evaluating $100 - 25\pi$, since $\pi = 3.142$. You can save time by approximating. By noting that π is approximately 3, $100 - 25\pi$ is slightly less than $100 - 75$ or 25. The only option less than 25 is (E).

8. Let the number be y
    then 12.5% of y = 40

    Thus $y = \dfrac{40}{12.5\%} = \dfrac{40}{12.5} \times 100 \dfrac{4000}{12.5} = 320$

    The correct answer is (D).

    Short cut: You are required to be familiar with some very basic fractions including eighths.
    Remembering that 12.5% is $\dfrac{1}{8}$, the problem can be rewritten:

    $\dfrac{1}{8}$ of y = 40

    Thus y = 40 × 8 = 320.

9. $\dfrac{351}{558}$ is equal to $\dfrac{39}{62}$, dividing through by 9 which is a common factor.

    The correct answer is (C).

    Short cut: It may not be obvious to you that 9 is a factor of both 351 and 558. In cases like this you may have to try dividing through by the first few smallest positive integers: 2, 3, 5, 7.

    In this instance 2 is not appropriate because 351 is odd. So try 3 and it works, giving $\dfrac{117}{186}$ (this incidentally tells you option (B) cannot be right).

    Try with 3 again (2 will still not be appropriate since 117 is odd). Dividing $\dfrac{117}{186}$ through by 3 yields $\dfrac{39}{62}$ which is option (C).

    Alternatively, take note that $\dfrac{351}{558}$ is very close to $\dfrac{350}{550}$ or $\dfrac{35}{55}$ or $\dfrac{7}{11}$ which is much higher than half. Now take a close look at the 5 options, (E) and (D) are less than half and can be eliminated. Option (B) is too close to one-half. Thus the choice is between options (A) and (C). At this stage divide 351 by the numerators of (A) and (C) and 39 comes out as a factor of 351 and not 109.

10. This is a very simple 'proportion' problem, but can be tricky and must be tackled properly.

500 km $\longrightarrow$ attracts \$80

Thus 1 km $\longrightarrow$ $\dfrac{\$80}{500}$

and y km $\longrightarrow$ $\dfrac{\$80}{500} \times y$ or $\dfrac{\$8y}{50}$

The correct answer is (B).

11. Put this information in the form of an equation

$$\frac{x}{21} - 1 = \frac{y}{21}$$

$$\frac{x}{21} = \frac{y}{21} + 1$$

This is the same as $x = y + 21$ (multiplying through by 21) and $y = x - 21$ (subtracting 21 from both sides).

The correct answer is (D).

12. By definition an odd number is one which is **indivisible** by 2. A close examination of the options reveals that in (B) we have an odd number n divided by an even number, m. Since m has a factor of 2 in it, we have an odd number divided by a number which may be written as $2 \times k$, where k is odd or even. In other words, we may write $\dfrac{n}{m}$ as $\dfrac{n}{2k}$, where $m = 2k$ (even).

Since n (odd number) cannot be divided by 2 without a remainder, the ratio n/m is not an integer. In all the other cases, the ratios **could be** an integer.

The correct answer is (B).

13. Let the man be x years old now. His son is therefore $\frac{x}{3}$ years old now. But the man had his son when he was 24.
Thus his current age (x) is the same as 24 plus the age of his son i.e.

$24 + \frac{x}{3} = x$ (i)
$72 + x = 3x$ (multiplying by 3)
$72 = 2x$ (subtracting x)
$36 = x$

The man is 36 years old. **The correct answer is (B).**

Short cut: This is a problem which while being simple can be confusing. Thus it may not be easy to derive equation (1) which is the key to the right answer. It is possible, and may indeed be faster, to get to the answer by using the method of substitution. The smallest and highest answers are 30 and 48. Try substituting these into the problem. If the man is 30 now, then his son is 6 and 6 is not a third of 30. So 30 is not right. If the man is 48, then the son is 24 now and 24 is not a third of 48. Try 36. If the man is 36, then his son is 36 − 24 = 12 and 12 is a third of 36, right. It may appear lengthy, but if you use this method in your preparation you develop a very fast mental analytical ability, which will be helpful in the actual examination.

14. This is an algebraic problem that requires a careful derivation of a set of simple equations which are subsequently solved.
We are required to determine the number of packets of beef-flavoured crisps.
Let b, v and p represent the number of packets of beef, vinegar and prawn crisps.
Then $b + v + p = 21$ .. (1) (given)
Also $p = 2v$ .. (2) (given)

The total cost of the crisps, 320 cents, is equal to the sum of the costs of the individual packets of crisps.
Thus $320 = 10p + 15v + 20b$ .. (3)

Substituting (2) into (1) we have
$b + v + 2v = 21$ or $b + 3v = 21$ .. (4).
Substituting (2) into (3), we have
$320 = 10(2v) + 15v + 20b$
or $320 = 35v + 20b$ .. (5)

Equations (4) and (5) are two equations with two unknowns, v and b, which can be solved for b or v.

320 = 35v + 20b (5)
21 = 3v + b (4)

Multiplying (4) by 20 and subtracting (5) from (4), we have

−320 = −35v − 20b
420 = 60v + 20b
100 = 25v
or v = 5
and from (4) b = 9

The correct answer is (E).

Short cut: Having gone through the analysis above you can see that it has been a very tedious and time consuming exercise! While one or two of the steps can be avoided or ignored, it still requires several minutes of analysis. A shorter (or smarter) way is by the method of substituting options into the problem. The facts provided are the total number of packets of crisps, 21, the individual unit prices and the total cost of all 21 packets of crisps, 320 cents. It is also given that there are 2 times as many packets of prawn crisps as vinegar. Try the smallest answer, option (D), if b = 7, then p + v = 14, and 14 cannot be divided between p and v in the ratio of 2 to 1. The same situation applies to option (C) and (B).

Options (A) and (E) however fit the data regarding numbers of packets of crisps.
For (A), if b = 12, then p + v = 9 and p and v can be 6 and 3 respectively.
For (E), if b = 9, then p + v = 12 and p and v can be 8 and 4 respectively.
But (A) does not fit the data with respect to cost i.e.
12 × 20 cents + 6 × 10 cents + 3 × 15 cents ≠ 320.

15. The first set of information implies that

    $n = 15k + 5$ .. (1)

    The second set of information implies that

    $n = 19m + 12$ .. (2)

    Since equations (1) and (2) are equal to each other, we have,

    $15k + 5 = 19m + 12$
    $15k = 19m + 7$
    or $15k - 19m = 7$

    The correct answer is (D).

16. If R = 25 sq units, then its side is 5 units. Similarly the lengths of sides Q and P are 3 and 4. Thus S is a right-angled triangle since the sides 5, 4 and 3 form a Pythagorean triad.

    The area of S is thus $\frac{1}{2} \times bh = \frac{1}{2} \times 4 \times 3$ or 6.

    The correct answer is (B).

ANALYSIS OF SOLUTIONS
Section 2
Problem Solving

1. $x = \dfrac{3}{2}\left(27 \div \dfrac{9}{8}\right)$

    $= \dfrac{3}{2}\left(27 \div \dfrac{8}{9}\right) = \dfrac{3}{2}(3 \times 8) = 3(3 \times 4) = 36$

    The correct answer is (D).

2. This problem requires us to investigate all the options one by one until we reach the correct option.

    Now suppose y is an even integer
    Then (A): $3y + 2$ is even
          (B): $9y$ is even
          (C): $4y + 5$ is odd, since $4y$ is even
          (D): y^2 is even
          (E): y^3 is even

    If y is an odd integer
    then (A): $3y + 2$ is odd (3 is odd too)
          (B): $9y$ is odd (9 is odd too)
          (C): $4y + 5$ is odd (5 makes the total odd since $4y$ is even)
          (D): y^2 is odd
          (E): y^3 is odd

    So whether y is even or odd (C) is always odd.
    The correct answer is (C).

    Short cut: You do not need to undertake all this time consuming writing. Just assume a y value of 3 (odd) and 2 (even) and check through the options.

3. Actual diameter $= \dfrac{0.5}{2000} = \dfrac{50}{2 \times 10^5} = 25 \times 10^{-5}$

    The correct answer is (A).

4. $2x^2 + 5x - q = 0$
when $x = -3$ the left hand side of the equation is
$2x^2 + 5x - q = 2(-3)^2 + 5(-3) - q$
$= 3 - q$
But this equals zero (right hand side)
∴ $3 - q = 0$ giving $q = 3$.

The correct answer is (C).

5. The required answer is a proportion or ratio of volumes.
Let V_1 = volume when diameter is 10ft,
and V_2 = volume when diameter is 12ft.

The volume of a sphere is given by $\frac{4}{3}\pi r^3$, where r is the radius.

Thus $V_1 = \frac{4}{3}\pi \times 10^3$

and $V_2 = \frac{4}{3}\pi \times 12^3$

The increase in volume is $V_2 - V_1 = \frac{4}{3}\pi (12^3 - 10^3)$

The ratio of the increase in volume to the initial volume is given by

$$\frac{V_2 - V_1}{V_1}$$

or $\frac{4}{3}\pi (12^3 - 10^3) / \left(\frac{4}{3}\pi \times 10^3\right) = \frac{12^3 - 10^3}{10^3}$

$= \frac{1728 - 1000}{1000} = 0.728$ or 72.8%

The correct answer is (D).

6.   B gets $\dfrac{10}{15+10+12} \times 24{,}300$

or $\dfrac{10}{27} \times 24{,}300$

or $9000

The correct answer is (C).

Short cut: Note that $\dfrac{10}{27}$ is nearly $\dfrac{9}{27}$ or $\dfrac{1}{3}$. So that $\dfrac{10}{27} \times 24{,}300$ is slightly more than $\dfrac{1}{3}$ of 24,300 or 8,100. Looking at the options only (C) can be right.

7.   Total weight of cars = 9 × 1.2 tonnes
     = 10.8 tonnes
     Distance covered = x km
     Total tonne – km = 10.8x

If cost per tonne – km is y.
Then cost of 10.8x tonne – km = 10.8xy$

The correct answer is (A).

8.   Unit cost of 1,000 tablet pack = $\dfrac{1200}{1000}$ cents = 1.2 cents

Unit cost of 500 tablet pack = $\dfrac{700}{500}$ cents = 1.4 cents

A tablet in the larger pack is 1.4 − 1.2 or 0.2 cents cheaper.

The correct answer is (D).

9. The term with the greatest value is the one with the smallest denominator since all the numerators are the same (=1). All the denominators have n^2, except option (E) and since $n>1$, option (E) has the smallest denominator and therefore the greatest value.

   The correct answer is (E).

10. Put the information down as clearly and as simply as you can. Let x be the number of nurses in the hospital.

    Then, $0.7x$ are female
    Therefore $0.3x$ are male
    0.6 of female nurses wear blue uniforms but the total number of female nurses is $0.7x$.

    Therefore $0.6 \times 0.7x$ female nurses wear blue uniforms. Total number of nurses who wear blue uniforms is $0.5x$ (half of x).

    Thus number of male nurses in blue uniforms = total in blue uniforms ($0.5x$) minus female nurses in blue uniforms ($0.42x$) = $0.08x$.

    Male nurses who do not wear blue uniforms = total male nurses ($0.3x$) minus male nurses who wear blue uniforms ($0.08x$).
    = $0.22x$ or 44% of the total number of nurses who do not wear blue uniforms.

    The correct answer is (C).

11. Since the base of the tank is a square, its width and length are equal. Let this be x in. We are also told that the height of the water in the tank is three times the width or $3x$.

    Volume = length × width × height
    = $x \times x \times 3x$
    = $3x^3$

    But this is equal to 81in^3. Hence $3x^3 = 81$ and $x = 3$ in and the square base has $3x^3$ or 9 sq inches of area.

    The correct answer is (C).

12. $\sqrt{32x}$ is the same as $\sqrt{16 \times 2x}$
or $\sqrt{16} \times \sqrt{2x}$
or $4\sqrt{2x}$

The correct answer is (B).

13. Draw the figure. A good diagram will make your job much easier.

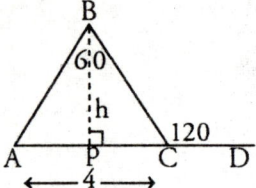

Area of a triangle is $\frac{1}{2} \times b \times h$. The base is given (4 ft). We need to find the height h.
$\angle ACB = 180 - 120 = 60$.
Therefore $\angle BAC = 60$ and $\triangle BAC$ is equilateral.
Therefore $AB = BC = AC = 4$ ft
and $AP = PC = 2$ ft.

Using the Theorem of Pythagoras:

$AB^2 = AP^2 + h^2$
$4^2 = 2^2 + h^2$
$h^2 = 16 - 4$
$h = \sqrt{12} = \sqrt{4 \times 3}$
$= 2\sqrt{3}$

Area of $\triangle BAC$ = 2 × area of BAP

$2 \times \frac{1}{2} \times 2 \times 2\sqrt{3}$

$= 4\sqrt{3}$ sq ft.

The correct answer is (D).

14. The best way to solve this kind of mixture problem is to organise the information in the form of a table. Let the volume of pure salt required be x litres.

	Vol	% brine	= solution
Original	100	20	100 (0.2)
Added	x	100	x(1)
Result	100+x	30	(100+x)(0.3)

From the last column
100 (0.2) + x (1) = (100 + x) (0.3)
20 + x = 30 + 0.3x
200 + 10x = 300 + 3x
7x = 100
x = 14.3

The correct answer is (E).

Short cut: Mixture problems are among the most difficult on the GMAT and are really extra-credit items. As the above solution shows, they are demanding in their analysis and also in terms of time. All the same, it is possible to develop a skill, determining the right option by using critical mental analysis of the problem by the method of substitution.

Let us choose a volume/quantity of pure salt that must be added to the solution. It must be pointed out that this choice should be based on good judgement, otherwise the whole trial and error method becomes very time consuming.

Since the original 100 litre solution is 20% salt, it means it is 80 litres water and 20 litres salt. So let us choose to try adding 20 litres of pure salt. We get a solution of 120 litre volume and 40 litres salt and this is 40/120 or 33% brine, which is higher than what is required. In other words, the volume of salt to be added must be less than 20 litres! Options (A) (B) and (C) can therefore be eliminated straight away, leaving options (D) and (E). At this point try substituting (D) or (E) into the problem and the right option comes out as (E).

15. Let the price per unit of the product be p. Let the total cost of producing n units be c.
Then if he sells n, the sales will be np.
But np = c + 0.2c = 1.2c (1)
If he sells only 90% of n, his sales will be 0.9 np and the ratio of this to total cost of producing n units is $\frac{0.9}{c}$ np

From (1), p = $\frac{1.2c}{n}$

putting this into (2)

Ratio = $\frac{0.9n}{c}$ X $\frac{1.2c}{n}$ = 0.9 X 1.2 or 1.08

The total revenue is 1.08 times or 8% greater than total costs.

The correct answer is (E).

Short cut: Do notice that if a sale of all n units leads to a revenue 20% higher than total costs (in fact a profit margin of 20%) then the profit margin will be less than 20% of total costs if only 90% of sales is done but the total cost of production remains the same. Profit equals sales minus costs. Thus the right answer will be less than 20%, which effectively eliminates options (A) and (B). At this point try substituting the smallest of the remaining options into the problem. Use the method of simple proportion as follows.

If 100% sales yield 120% of costs then 90% sales will yield $\frac{90}{100}$ X 120% of costs which equals 108% or 8% over costs.

16. All the options except (1) have a coefficient which is even thereby making the terms even. Only (1) has an odd coefficient, thereby making 9S odd if S is odd.

The correct answer is (D).

ANALYSIS OF SOLUTIONS

Section 3

Data Sufficiency

1. Statement (1) is sufficient. With annual rates of growth we can calculate the price year after year to the third year.

   Year 2 price = Year 1 price plus 10%
                 = $8000 + 0.1 (8000)
                 = $8800

   Year 3 price = Year 2 price plus 10%
                 = $8800 + 0.1 (8800)
                 = $9600

   Statement (2) is also sufficient. The price in 2 years time will be 1.2 x $8000 = $9680.

   The correct answer is (D).

2. Statement (1) is not sufficient. From $y + z = 4$, we can derive $(y - z) = (4 - 2z)$ (subtracting 2z from both sides). But that is all we can do and we still have not evaluated $(y-z)$.

   Statement (2) is also not sufficient.
   $y-z < 4$, means that $(y - z)$ can be 3, 0, –10, – 100 etc.
   i.e. there is no unique value of $(y - z)$.

   If we combine the two statements, we end up with $(4 - 2z) < 4$ or $(2 - z) < 2$ or $z>0$, which is also not unique. Thus more information is required and the correct answer is (E).

3. **Statement (1)** is not sufficient. We are given the original price but not the final. We therefore cannot calculate the reduction in price.

   Statement (2) is also not sufficient. We know that the absolute price reduction is $13 but we do not know the original price so we cannot calculate the fraction, which must be 13 divided by the original price.

   But if we combine the two statements then we have both the level of reduction, $13, and the original price, $260, giving us $\frac{13}{260}$ or $\frac{1}{20}$

   The correct answer is (C).

4. **Statement (1)** gives a ratio only and not actual measures. We need an actual measure, so this statement is not helpful.

   Statement (2) gives us the perimeter of the rectangle, $2L + 2W$, but we do not know the components L and W. Statement (2) alone does not help. Combining statements (1) and (2) we can form 2 equations as follows:

   $L : W = 3:2$
   or $\frac{L}{W} = \frac{3}{2}$
   or $L = \frac{3}{2}W$ (i)
   Also $2L + 2W = 192$ (ii)

   From these we can find the values of L and/or W.

   The correct answer is (C).

5. **Statement (1)** is woefully inadequate. The 3 numbers greater than 10 can be 16, 18, 20 or 30, 50, 100 or 9000, 13453, 9m etc. Indeed there is very little in statement (1).

Statement (2) is also inadequate. Between 5 and 20, there are 6, 7, 8, 9, 10, ... 19; in fact 14 numbers! Which 6 are we concerned with? Even if we combine both statements, we still do not have a unique set of 6 numbers, the mean of which we need to calculate.

The correct answer is (E).

6. **Statement (1)** is sufficient. If $a = b$, then $a^2 = ab$.

Statement (2) is a mere identity and does not contain anything about the values of a and b.

The correct answer is (A).

7. **Statement (1)** is not sufficient. If PSR is obtuse, it means that PQR is also obtuse.

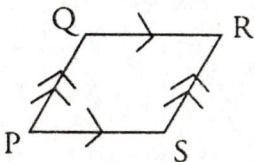

But an obtuse angle is one greater than 90, but less than 180 degrees and can be 179, 160, 145, 95; in fact millions of angular measures lie between 90° and 180°.

Statement (2) says that angle SPQ is 70°. This implies that PQR must be 110 degrees, since the two must sum to 180° degrees.

The correct answer is (B).

8. This is a right-angled triangle with ∠ B=90° and it is one of the two mentioned in Section 1 that you should be familiar with. In this particular right triangle, the sides are in proportion as shown in the diagram. If we know the length of one side, we can calculate the length of the other two sides.

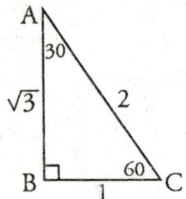

Statement(1) provides the length of one side, the hypotenuse AC, as 13. So we can calculate the length of side BC and statement (1) is sufficient.

Statement (2) also provides the length of one side and is also sufficient.

The correct answer is (D).

9. Statement (1) is not sufficient. If $\frac{n}{4}$ is a positive integer, this implies that $\frac{n}{4}$ = 1, 2, 3, 4... The values of n are thus 4 ($\frac{n}{4}$ = 1), 8 ($\frac{n}{4}$ = 2), 12 ($\frac{n}{4}$ = 3) etc. While n = 4 is less than 5, all the other values of n are greater than 5. So statement (1) does not lead to a single answer.

Statement (2) is sufficient. If $\frac{n}{6}$ is a positive integer then $\frac{n}{6} > 0$ and $\frac{n}{6}$ = 1, 2, 3, 4, 5, ...
Taking the minimum value 1,
$\frac{n}{6}$ = 1 means n = 6.

In all cases n must be an integer, the lowest being 6.

The correct answer is (B).

10. Statement (1) is sufficient. With Q = 1000, we can compute the cost functions for both plants and compare them.

    Statement (2) says nothing about costs and is not sufficient. It is tempting to think that because plant B is newer it will have lower costs, but nobody has said so.

    The correct answer is (A).

11. Statement (1) is not sufficient. There is nothing to suggest that all those who actually buy the product receive gifts.

    Statement (2) is sufficient. If the number of people who take part in the exercise is x, say, then $0.5x$ promise to buy and 0.2 of $(0.5x)$ actually buy.

    The percentage of those who buy is
    $$\frac{0.2\,(0.5x)}{x} \times 100 = 10 \text{ percent.}$$

    The correct answer is (B).

12. Statement (1) is tricky but sufficient.
    $x > r$ is equivalent to $x - r > 0$ (positive)
    and $y < s$ is equivalent to $y - s < 0$ (negative).

    Now $r - s > x - y$ is equivalent to $(y - s) > (x - r)$, moving y to the left hand side and moving r to the right hand side.

    The question we need to answer, 'is $r-s > x - y$?' can be put in the form is '$(y - s) > (x - r)$?'

    From the information provided in statement (1) $(y - s)$ is negative and $(x - r)$ is positive and the question reduces to:

    Is (negative number) > (positive number)?
    The answer is clearly no.

    Statement (2) provides values for all the variables so we can determine whether $(6 - 8) > (7 - 4)$ or not.

    The correct answer is (D).

13. **Statement (1) is not sufficient.** We are only provided with relative speed, not actual speed and time taken. We cannot determine the distances covered by Nigel and Freydis, nor can we determine the difference in the distances covered.

    Statement (2) is also not sufficient. We have times but no speeds so we cannot determine distances, let alone the differences. Now let us combine the two statements, we have the time periods and relative speed. There is no information on absolute speeds, so we cannot calculate distances. Further information is required and **(E) is the correct option.**

14. Speed = distance/time
    Statement (1) is sufficient as it provides the distance. (Time is provided in the problem).

    Statement (2) is not sufficient. We do not know for how long the car travelled at 35 km/hr or what the maximum speed was.

    The correct answer is (A).

15. **Statement (1)** implies that y = 3. Since we are only interested in positive numbers the answer must be 5 and so statement (1) is sufficient.

    Statement (2) is tricky but becomes easy if you remember that when you invert an inequality the direction of the inequality is reversed. For example 4 > 3 when inverted becomes
    $$\frac{1}{4} < \frac{1}{3}$$

    As you can see the inequality is reversed.

    Statement (2) states that
    $$\frac{1}{4} < \frac{1}{y} < \frac{1}{2}$$
    which is equivalent to
    4 > y > 2

    Since y is a positive integer it can only be equal to 3 and statement (2) is sufficient.
    The correct answer is (D).

16. Statement (1) is sufficient.

$$\text{Percent profit} = \frac{\text{selling price} - \text{cost price}}{\text{cost price}} \times 100$$

Statement (1) provides the selling price of $1200. The percent profit is 40 and we can deduce the cost price from the above equation ($857).

Statement (2) is not sufficient. There is no information on the absolute value of the selling price.

The correct answer is (A).

17. n has a value between 1 and 9. Therefore N = 421−n means N has a value between (421 − 1) and 421 − 9 or 412 and 420.

Using information contained in **statement (1)** 413 and 420 are the two numbers divisible by 7 in the range 412 to 420. If N is 413 then n must be 8 and if N is 420, then n must be 1. We have two values of n and statement (1) does not give us a unique answer.

Statement (2) implies that N may be 414, 417 and 420 (all between 412 and 420 and divisible by 3). These also lead to 3 values of N and this is therefore not sufficient.

Combining both statements, however we obtain N=420, which is common between statements and (2) leads to a value of 1 for n.

The correct answer is (C).

18. ABCD is a parallelogram if $\angle A = \angle C$ and $\angle D = \angle B$.
Statement (1) provides that $\angle A = \angle C$ but says nothing else. Thus the statement is not sufficient.

Statement (2) says nothing about AD and BC and is not sufficient.
Combining both statements, however, we note that $\angle C + \angle B = 180$ and $\angle B + \angle D = 180$. Also $\angle A = \angle C$. Thus AD must be parallel to BC and the figure ABCD is thus a parallelogram.

The correct answer is (C).

19. It is absolutely essential to take note of two things.
First, that there is nothing to suggest that a and b are positive numbers.
Second, that $a^2 - b^2 = (a + b)(a - b)$.

Statement (1) states that in $(a + b)(a - b)$, the second bracket contains a positive number. We cannot tell if the first bracket, $(a + b)$, is also a positive number. Let us examine it. If for example $a = 4$ and $b = -6$, the second bracket is $4-(-6)$ or 10 which is positive and is in line with statement (1).

But the first bracket is $(4 - 6)$ or (-2) which is negative, thereby making the whole of $(a + b)(a - b)$ negative. On the other hand if both a and b are positive with b<a, then statement (1) is satisfied and the whole of $(a + b)(a - b)$ is then positive. In fact to satisfy statement (1) a and b can both be negative but with $a < b$, making $(a + b)(a - b)$ negative. Thus statement (1) does not contain enough information for us to determine whether or not $(a + b)(a - b)$ is positive.

Statement (2) does not say anything about $(a - b)$. If for example $a = 2$ and $b = 4$ then $(a + b)$ is positive (satisfies statement (2)) but $(a - b)$ is negative, thereby making $(a+b)(a-b)$ negative. On the other hand if $a = 4$ and $b = 3$, then $(a + b)(a - b)$ is positive. There is no unique answer to the question and statement (2) is not sufficient.

By combining the two statements, however, we know (a–b) is positive (from statement (1)) and $(a + b)$ is positive from statement (2) and the product of two positive numbers is positive, a definite answer.

The correct answer is (C).

20. From statement (1) we can deduce that $N = 5K + 1$, where K is the whole number obtained when N is divided by 5. When K is odd, N is even but when K is even, N is odd. Statement (1) is therefore not sufficient.

Statement (2) is, however, sufficient. Since we can write $N = 8y + 1$, N is odd regardless of whether y is odd or even.

The correct answer is (B).

ANALYSIS OF SOLUTIONS
Section 4
Data Sufficiency

1. Statement (1) alone is not sufficient since we do not know what has happened to the width. Statement (2) alone is not sufficient for a similar reason. However, by combining the two statements we can reach a solution.

    Let the initial length and width of the rectangle be L and W. The area is LW square units.

    If the length and width are altered to 1.2L and 0.9W respectively, then the new area is 1.2L × 0.9W which equals 1.08WL. This is 8 per cent more than WL.

    The correct answer is (C).

2. If we can find the value of x then y can be determined. Statement (1) gives us $x = \frac{1}{2}y$. With this we can determine y. Thus statement (1) is sufficient.

    Statement (2) is a quadratic equation with solution $x = 2$. With this y can be evaluated and statement (2) is also sufficient.

    The correct answer is (D).

3. Statement (1) is not sufficient. It does not tell us anything about the cost of the next 40 copies.

    Statement (2), however, is sufficient. If it costs 10 cents a copy then for 50 copies it must cost 50 × 10 cents or $5.00.

    The correct answer is (B).

4. From **Statement (1)** we can work out that the distance is 50 mph × 5 hours or 250 miles.

   From **Statement (2)** we can calculate that the distance is 500 mph × $\frac{1}{2}$ hour, or 250 miles. Both statements are sufficient.

   The correct answer is (D).

5. Note that the question can be rephrased as 'is xy negative?' **Statement (1)** is sufficient. If the quotient of two numbers is negative, then they must have opposite signs. If they have opposite signs, then their product will also be negative.

   Statement (2) simply means that x is smaller than y, eg. $x = 5$ and y = 8, or that the absolute value of x is greater than that of y, if both of them are negative. Statement (2) does not lead to a simple conclusion and is therefore not sufficient.

   The correct answer is (A).

6. Neither statement (1) nor statement (2) is individually sufficient. **Statement (1)** only indicates a ratio. **Statement (2)** tells us nothing about factory x.

   But a combination of the two statements can give us a solution. If factory x turns out 1.5 times the number of cars that factory y produces, which is 10,000 then factory x turns out 15,000 cars.

   The correct answer is (C).

7. Clearly, neither of the two statements is individually sufficient. But combined, they lead to a solution.

   From statement (1), P>4 −m.

   From statement (2) the value of m must be less than 3.

   Even if we make m = 3, then statement (1) gives us P>1, and this is always true.

   The correct answer is (C).

8. This is a tricky question. Since the equation $2x + 2xy + z = 2xy + 6$ appears to have 3 unknowns (x, y, and z), you may be tempted to seek values for both y and z (using both statements) in order to solve the equation and find x. In this case you would choose option (D).

   However, a careful examination of the equation reveals that it simplifies to $2x + z = 6$

   Thus, only the value of z is required in order to find x and so only statement (1) is needed.

   The correct answer is (A).

9. Both statements are sufficient.

   Triangle OAB is isoceles. If y = 60 then x must be $\frac{120}{2}$ or 60.

   Statement (2) indicates that triangle OAB is equilateral. Thus x = 60.

   The correct answer is (D).

10. Both statements give ratios only, and no absolute values. They are not sufficient, either individually or combined. Further information is required.

   The correct answer is (E).

11. Statement (1) indicates that K is even. It is sufficient to answer the question.

   Statement (2) is not sufficient. K can be odd or even.

   The correct answer is (A).

12. Neither statement (1) nor statement (2) is individually sufficient. To work out the area we need to know the length of the base (which is given as 10) and the height, which we need to determine. Neither of the statements, considered individually, give us the height.

   However, by combining the statements we have an isosceles triangle with sides AB = AC = 6 and base 10. The height is $\sqrt{36 - 25}$ or $\sqrt{11}$.

   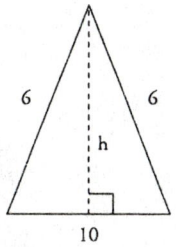

   The correct answer is (C).

13. **Statement (1) is sufficient.** $2x = 2y + 10$ can be written as $2(x - y) = 10$, or $(x - y) = 5$. Thus $(x - y)$ can be evaluated as 5.

    Statement (2) is not sufficient. If $x = 8$, then $(x - y)$ reduces to $(8 - y)$. We still don't know the solution.

    The correct answer is (A).

14. **Statement (1) is sufficient.** If the original area is 90 ft^2, we can calculate the diameter from the formula, $A = \dfrac{\pi d^2}{4}$

    By adding 2 ft to the original diameter we obtain the new diameter with which we can compute the new area. The original area can then be subtracted from the new area to determine the increase in area. The solution is as follows:

    $$\text{Old diameter} = \sqrt{\dfrac{\text{area} \times 4}{\pi}} = \sqrt{\dfrac{90 \times 4}{\pi}}$$
    $$= 10.7 \text{ ft}$$

    $$\text{New area} = \dfrac{\pi \times (12.7)^2}{4}$$
    $$= 126.77$$

    Increase in area = $126.77 - 90 = 36.8$

    Statement (2) is also sufficient. If the original diameter is 10.7ft, then the new diameter is 12.7ft and we can calculate the areas and subsequently the difference.

    The correct answer is (D).

15. Neither of the two statements is individually sufficient.

    In **Statement (1)**, if a number is divisible by 5 it does not necessarily mean that it is divisible by 15 – consider 25, 10, 40. The same argument applies to **Statement (2)**.

    However, a combination of the two statements yields a solution. If a number is divisible by both 3 and 5, then it means that 3 and 5 are factors of that number and therefore 3 X 5, or 15, is also a factor.

    The correct answer is (C).

16. **Statement (1)** is sufficient because of the theorem that states that a line like TS which divides two sides of a triangle in the same ratio must be parallel to the third side.

    Statement (2) is also sufficient. If angle PTS is equal to angle PQR, this means they are corresponding angles and so TS must be parallel to TR.

    The correct answer is (D).

17. **Statement (1)** is not sufficient since we do not know the signs of x and y.

    Statement (2) is not sufficient. It is true only for positive or absolute values. For example, if $y = 5$ and $x = 3$ ($y>x$), then $y^2 > x^2$. But if $y = -5$ and x -3 then $y < 3$, even though $y^2 > x^2$.

    The correct answer is (E).

18. Neither statement is sufficient as neither gives any absolute values. Further information is required.

    The correct answer is (E).

19. Statement (1) is not sufficient as we have no idea how many telephones are in the box.

    Statement (2) is not sufficient either as we are not told how many telephones there are in total.

    A combination of the two statements shows that the 30 telephones that are not blue represent 37.5% (100 − 62.5) of the total. The total must therefore be 80, and there must be 62.5% of 80 or 50 blue telephones in the box.

    The correct answer is (C).

20. Statement (1) is clearly insufficient as it is only a ratio.

    Statement (2), however, is sufficient. The sum of the first 8 positive integers is 1 + 2 + 3 + 4 + 5 + 6 + 7 + 8 = 36. One sixth of 36 = 6. Thus John is 6 years old.

    The correct answer is (B).

GMAT
Going for a high score?

Boost your score and confidence by attending our one-day intensive revision course

"Success at the GMAT"

Held throughout the year in central London.

- Learn from experienced professionals
- Practice on favourite exam topics
- Develop a successful examination technique
- Acquire a winning revision strategy
- Receive valuable course notes

Contact The MBA & GMAT Advice Centre on 01565 755226 for full details.

Essential Reading for All MBA Candidates

A specially selected range of books is available by mail order from the MBA & GMAT Advice Centre. We import the best titles from the USA and also publish our own GMAT revision books. All books are despatched on the day the order arrives.

Titles currently available include:

- **GMAT Practice Examinations**
- **The Official Guide for GMAT Review**
- **The Official Guide to MBA Programs**
- **The AMBA Guide to Business Schools**

Contact us by telephone or fax for immediate service:

**The MBA & GMAT Advice Centre, PasTest,
Egerton Court, Parkgate Estate, Knutsford,
Cheshire WA16 8DX
Telephone: 01565 755226 Fax: 01565 650264**